MAMMALIAN CELL CULTURE

ESSENTIAL TECHNIQUES

Edited by

A. Doyle and J. Bryan Griffiths

ECACC, Centre for Applied Microbiology and Research, Salisbury, UK

JOHN WILEY & SONS

Chichester • New York • Weinheim • Brisbane • Singapore • Toronto

Published in association with BIOS Scientific Publishers Limited

British Library Cataloguing in Publication Data
A catalogue record for this book is available from the British Library.

ISBN 0 471 97057 3

Typeset by Footnote Graphics Ltd, Warminster, UK
Printed and bound in UK by Biddles Ltd, Guildford, UK

ESSENTIAL TECHNIQUES SERIES

Series Editor
D. Rickwood
Department of Biological Sciences, University of Essex, Wivenhoe Park, Colchester, UK

Published titles
Antibody Applications
Gel Electrophoresis: Nucleic Acids
DNA Isolation and Sequencing
Gel Electrophoresis: Proteins
Gene Transcription: DNA Binding Proteins
Gene Transcription: RNA Analysis
PCR
Cell Biology
Human Chromosome Preparation
Antibody Production
Mammalian Cell Culture

Forthcoming titles
Plant Molecular Biology
Nucleic Acid Hybridization
Vectors: Cloning Applications
Vectors: Expression Systems

CONTENTS

CONTRIBUTORS

Al-Rubeai, M. School of Chemical Engineering, University of Birmingham, Edgbaston, Birmingham B15 2TT, UK

Azghani, A. Department of Biochemistry, University of Texas Health Center, PO Box 2003, Tyler, TX 75710, USA

Bolton, B. European Collection of Cell Cultures, Centre for Applied Microbiology and Research, Porton Down, Salisbury, Wilts SP4 0JG, UK

Brown, K. AFRC Institute of Animal Physiology and Genetics, Cambridge Research Station, Babraham, Cambridge CB2 4AT, UK

Chen, T.R. American Type Culture Collection, 12301 Parklawn Drive, Rockville, MD 20852, USA

Clarke, J.B. MFN Laboratories, Springwood Industrial Estate, Braintree, Essex, UK

Cour, I. American Type Culture Collection, Culture Department, 12301 Parklawn Drive, Rockville, MD 20852, USA

Doyle, A. European Collection of Cell Cultures, Centre for Applied Microbiology and Research, Porton Down, Salisbury, Wilts SP4 0JG, UK

Griffiths, B. Centre for Applied Microbiology and Research, Porton Down, Salisbury, Wilts SP4 0JG, UK

Jones, B.P.C. NIBSC, Blanche Lane, South Mimms, Potters Bar, Herts EN6 3QG, UK

MacDonald, C. Department of Biology, University of Paisley, Hugh Street, Paisley PA1 23E, UK

Morris, C. University of Oxford, The Wellcome Trust Centre for Human Genetics, Windmill Road, Oxford OX3 7BN, UK

Mowles, J.M. Biochem ImmunoSystems, 20 Woking Business Park, Albert Drive, Woking GU21 5JY, UK

Newell, D. Central Veterinary Laboratory, Biotechnology Unit, Newhaw, Weybridge KT15 3NB, UK

Simione, F. American Type Culture Collection, 12301 Parklawn Drive, Rockville, MD 20852, USA

Warburton, S. European Collection of Cell Cultures, Centre for Applied Microbiology and Research, Porton Down, Salisbury, Wilts SP4 0JG, UK

West, C.M.L. The Paterson Institute for Cancer Research, Department of Radiobiology, Wilmslow Road, Worthington, Manchester M20 9BX, UK

Wilson, A. Oncology Research Laboratory, Derby City General Hospital, Derby DE22 3NE, UK

Haynes, L. School of Biological Sciences, University of Bristol, Woodland Road, Bristol BS8 1UG, UK
Jenkins, N. School of Applied Science, De Montfort University, Leicester LE1 9BH, UK
Worthington, V. Worthington Biochemical Corporation, 450 Halls Mill Road, Freehold, NJ 07728-8806, USA

ABBREVIATIONS

CE	cloning efficiency
CFE	colony-forming efficiency
cfu	colony-forming unit
CHO	Chinese hamster ovary
COSHH	Control of Substances Hazardous to Health
DMEM	Dulbecco's modification of Eagle's medium
DMSO	dimethyl sulfoxide
EBV	Epstein–Barr virus
EDTA	ethylenediamine tetraacetic acid
ELISA	enzyme-linked immunosorbent assay
FBS	fetal bovine serum
FTM	fluid thioglycollate medium
GCFC	genomic chromosome fingerprinting characteristics
HAT	hypoxanthine–aminopterin–thymidine
HBSS	Hank's balanced salt solution
HEPA	high-efficiency particulate air
HEPES	4-(2-hydroxyethyl)piperazine-1-ethanesulfonic acid
LAF	laminar airflow
LDH	lactate dehydrogenase
MDCK	Madin–Darby canine kidney
MEL	maximum exposure level
MEM	minimal essential medium
MTT	3-(4,5-dimethylthiazol-2-yl)-2,5-diphenyltetrazolium bromide
PBS	phosphate-buffered saline
PCR	polymerase chain reaction
PD	population density
PE	plating efficiency
PEG	polyethylene glycol
PHA	phytohemagglutinin
QAC	quaternary ammonium compound
QC	quality control
TSB	tryptone soya broth
VNTR	variable number of tandem repeats

PREFACE

Cell culture is regarded as core expertise in a wide range of biological and medical science technological applications. It has a universality that may be misleading when it comes to the need to adhere to *accepted* procedures. Increasingly, standardization, indeed accreditation, are the norm, as a set of well documented procedures should be of value to all scientists using these techniques.

We have invited contributions from key individuals recognized in each field and have included our own personal choice of essential techniques. We would like to thank Janet Nuth who has provided valuable support in producing this volume.

Alan Doyle and J. Bryan Griffiths

SAFETY

Attention to safety aspects is an integral part of all laboratory procedures and national legislations impose legal requirements on those persons planning or carrying out such procedures. While the authors, editor and publisher believe that the recipes and practical procedures, as set forth in this book, are in accord with current recommendations and practice at the time of publication, they accept no legal responsibility for any errors or omissions, and make no warranty, expressed or implied, with respect to material contained herein. It remains the responsibility of the reader to ensure that the procedures which are followed are carried out in a safe manner and that all necessary safety instructions and national regulations are implemented.

In view of ongoing research, equipment modifications and changes in governmental regulations, the reader is urged to review and evaluate the information provided by the manufacturer, for each reagent, piece of equipment or device, for any changes in the instructions or usage and for added warnings and precautions.

These are freely accessible using:
http://physchem.ox.acuk/MSDS

Other Internet sites providing safety information are:
gopher://atlas.chem.utah.edu/11/MSDS
gopher://ecosys.drdr.virginia.EDU:70/11/library/gen/toxics
gopher://portfolio.stanford.edu.2270/1gopher%20SHOW INDEX%20TCHEM
http://hazard.com/msds/index.html
http://www.fisher1.com/msds.htm

Given the rapidly evolving nature of the Internet, the addresses of these sites may vary and new sites appear. The reader can check for new sites using one of the net search engines such as yahoo or infoseek.

You are actively encouraged to check these data sheets to confirm our assignments and for more detailed information on individual hazards; however the author, editor and publisher can accept no responsibility for any material contained in these data sheets. Furthermore, you must always follow the precautions outlined on labels and data sheets provided by individual manufacturers.

All procedures mentioned within this book must be carried out under conditions of good laboratory practice in accordance with local and national guidelines. Some procedures involve specific hazards, including but not limited to hazards in the following categories:

Chemical. A number of the reagents are known to be carcinogenic, mutagenic, toxic, inflammable, highly reactive or otherwise hazardous. Substances known to be hazardous have been marked with the symbol ⚠ in the list of reagents (but not subsequently) for each protocol, or if they appear as alternatives to the main protocol, the *first time* they appear in the notes. The reader should consult the safety notes on these pages before embarking on any of the procedures covered. This is in no way meant to imply that undesignated chemicals are nonhazardous, and all laboratory chemicals should be handled with extreme caution. Information is not available on the possible hazards of many compounds. The criteria we have generally used for denoting a substance with ⚠ is based upon a hazard level of 2 or more (on a scale of 0–4) in any of the categories in the Baker Saf-T-Data™ system used in the material safety data sheets (MSDS) held at the University of Oxford, UK.

Radiation. The use of radioisotopes is subject to legislation and requires permission in most countries. Furthermore, national guidelines for their use and disposal must be rigorously adhered to. The procedures in protocols that use radioisotopes must only be carried out by individuals who have received training in the use of such material using the appropriate facilities, protection and personal monitoring procedures.

Biological. Antibodies, sera and cells (particularly, but not exclusively, those of human and nonhuman primate origin) pose a significant biological hazard. All such materials, whatever their origin, may harbor human pathogens and should be handled as potentially infectious material in accordance with local guidelines. Any recombinant DNA work associated with protocols is likely to require permission from the relevant regulatory body and you must consult your local safety officer before embarking upon this work.

Electrical. Many of the procedures in this book use electrical equipment. Electrophoresis techniques may present particular hazards of this nature.

Lasers. Flow cytometers and certain other types of laboratory equipment contain lasers. Users should ensure they are fully aware of the potential hazards of using such equipment.

I INTRODUCTION. A. Doyle, C.B. Morris and B. Jones

Laboratory facilities and tissue culture technique. A. Doyle and C.B. Morris

The tissue culture laboratory

One characteristic of cell culture technology is that it generally post-dates the laboratory facilities in which it is carried out. In most cases the adaptation of existing facilities has taken place, at best, with a restricted budget and a low priority. This introductory chapter is offered with two concepts in mind. The first is to give sufficient information and background knowledge to design the ideal facility for tissue culture. The second is to provide a minimum set of criteria for conducting tissue culture procedures in a competent, efficient and safe manner without necessarily requiring the kind of investment that ideal facilities would require.

There are several aspects to the good design of tissue culture laboratories, as described below.

Single-use facilities: It is essential to have procedures which allow for the handling of newly acquired cell lines, either developed from primary sources or obtained from elsewhere, under a restricted 'quarantine' basis. This can be achieved by having an isolated laboratory under negative pressure, high-efficiency particulate air (HEPA)-filtered air control systems with dedicated work stations (e.g. class II containment cabinet) and protocols which restrict the movement of staff and equipment into and out of this area.

Figure 1 shows the work areas in an idealized tissue culture facility. Positive-pressure HEPA-filtered air systems (A) are used for sterile procedures, preparation and dispensing of media and other supplies, filtration of media components, preparation of microbiological quality control reagents and preparation of sterile equipment. Cell lines should not be handled in this area.

The main tissue culture facility (B) may be under positive or negative HEPA-filtered air systems, depending upon the nature of the material being handled. Basic recommendations would be to handle under positive pressure only human diploid cells

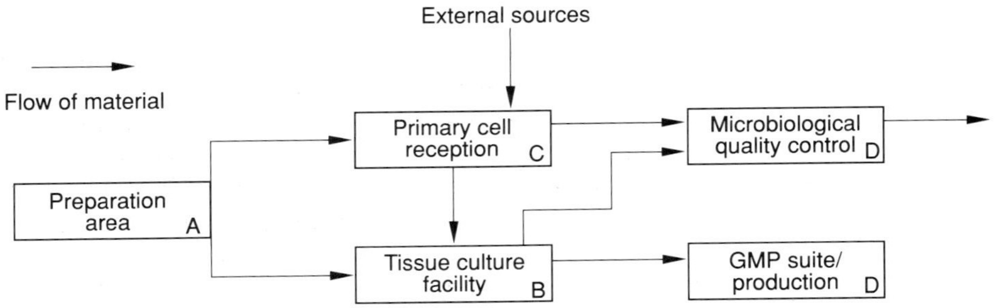

Figure 1. Flow diagram of work areas in an idealized tissue culture facility.

and other well-characterized cell lines, with extensive quality control performed to rule out the presence of any adventitious agents.

All incoming material from primary sources or other laboratories must be handled separately from other material routinely tested under existing quality control procedures, in the primary cell reception (C). Once testing has been completed, the cells can be transferred to the main tissue culture facilities.

Other designated work areas (D) should be assigned to specific tasks if possible. An essential requirement of good manufacturing practice cell banking is a dedicated, controlled and validated working environment. Research and development production and other procedures with cell lines should be conducted elsewhere from the routine maintenance of cell stocks. Movement of staff, materials and equipment, and disposal of items from these areas, should be controlled and monitored. All equipment should be regularly serviced and any HEPA filters checked regularly. This could vary from full-scale filter testing to the routine use of microbiological settle plates to monitor the level of contamination. In a well-designed facility it should be the norm to conduct all tissue culture procedures in the absence of antibiotic cover.

Multi-use facilities: Multi-purpose laboratories are difficult to operate and sensible practices may lead to some restriction on activities. Under such circumstances greater planning and foresight are necessary. Work schedules need to be planned with a flow-through of material on a day-to-day or single-day basis. Preparation work, handling quality controlled cell lines, handling 'in process' quality control cell lines and thorough cleaning of 'work areas' is required between activities. It is absolutely imperative to achieve a good laboratory routine in shared facilities. Again, class II containment cabinets provide good working facilities if operated correctly and checked periodically for efficiency. A particular concern is the mycoplasma contamination status of any incoming material.

Tissue culture technique

Prevention of contamination by any extraneous organisms (bacteria, yeast, fungi, mycoplasma, viruses) is a prerequisite in tissue culture. The basic principles of good technique are essentially common sense.

Blanket antibiotic cover is to be avoided at all cost. Indiscriminate antibiotic use will inevitably lead to problems. Hidden infections can occur, which only become apparent during experimental or production procedures, and may render the use of the cell line impossible for that particular purpose.

The basic guidelines for good laboratory technique are as follows:

(i) Avoid multi-use facilities if possible.
(ii) Only handle one cell line at a time. Such a common-sense approach should cut to the minimum the possibility of mislabeling, mix-ups and cross-contamination between cell lines.
(iii) Examine cultures daily for evidence of gross-contamination.
(iv) Maintain separate bottles of media (and preferably other reagents) for each cell line.
(v) Quality control all of the reagents used in tissue culture, even if bought in from a commercial supplier.
(vi) If using class II containment facilities, keep the work areas uncluttered (consumable supplies should be kept in racks below the work area), swab with a suitable disinfectant between handling cell lines and allow a minimum of 15 min between the handling of different cell lines.

(vii) Dispose of all used materials safely, efficiently and routinely.
(viii) If separate work areas are not available, at least designate separate incubators for different types of materials.
(ix) Routine quality control procedures should be carried out more frequently in multiple-use facilities.
(x) Ensure that laboratory coats are changed regularly.
(xi) Wash hands thoroughly before and after commencing laboratory work. If possible use sterile gloves.
(xii) Avoid the generation of aerosols (e.g. by pipetting too vigorously) in the work area.
(xiii) Liberal use of disinfectant (e.g. 70% ethanol) to swab work surfaces/bottles/flasks is recommended.
(xiv) Train and monitor junior staff in tissue culture technique before allowing them to handle particularly valuable material.
(xv) Maintain facilities efficiently. Test filters routinely and, if manometers are fitted, ensure that a reporting procedure is instituted for changes that may indicate system failure.
(xvi) Ensure that all incubators, microscopes, centrifuges and other equipment is cleaned and serviced regularly. Humidified incubators require special routine cleaning procedures to avoid contamination.
(xvii) Discard unused cultures from incubators as soon as they are no longer required. If laboratory topography and finance permits, install facilities which can be routinely fumigated.

Safety aspects (handling, equipment and containment). B. Jones

Guidelines issued by national regulatory bodies attempt to set realistic safety standards for all work related to the handling of biological agents, including cell and tissue cultures. It is now generally accepted that biological agents can be classified into four hazard groups according to their nature and that for each group there is an appropriate containment level under which such organisms should be handled. Uncontaminated animal cell cultures do not normally present a biohazard to scientific staff or the environment but they are proliferating systems and can sustain the survival or replication of a range of adventitious agents.

Which containment levels are to be used will depend to some extent on the nature and type of work being undertaken. If the work involves cells or organisms that are unlikely to cause human disease, then little more than good laboratory practice and discipline will be required. If, however, cells or organisms can cause human disease, then other containment levels will need to be used. The extent to which the individual levels are used will depend upon the individual risk assessment of the work being undertaken.

Risk assessments

The history of animal cell culture has demonstrated that the cells can be handled safely, yet it is essential that they are included in any risk assessments. To simplify the risk assessment, cell cultures may be allocated to one of several groups as shown in *Table 1* [1, 2].

The true identity of the cells cannot always be assumed as there are numerous reports of cross-contamination. It is important, therefore, to use only those cell strains and cell lines that have been authenticated. These are best obtained from the originator of the cell culture or a recognized cell culture collection.

Table 1. Cell culture groups

Hazard	Cell type	Containment
Low: uncertain	Well characterized/authenticated finite cell lines of human or primate origin. Nonhuman, nonprimate cell lines which have been authenticated, have a low risk of endogenous infection with a human pathogen and present no apparent hazard to laboratory workers	Containment level 2 and use of safety cabinet
Medium: uncertain	Cell lines/strains not fully authenticated or characterized	Containment level 2 and use of safety cabinet
High: defined	Cells with endogenous pathogens and cells deliberately infected	Containment appropriate to the pathogen
High: uncertain	Primary cells from blood, lymphoid cells, neural tissue of human and simian origin	Containment appropriate to the potential risk

Methods available

Laboratory practice and discipline

Besides the physical aspects of laboratory containment, systems such as laboratory practice and discipline, bench technique, occupational safety and hygiene, industrial large-scale practice, effective disinfection and sterilization procedures will all provide good primary containment measures in any laboratory [1–5]. In many countries there is a legal requirement for written systems of work laid down in local rules, codes of practice, hazard assessments [6], and standard operating procedures that will cover specific procedures and laboratories.

Transportation

Materials such as specimens, cell cultures and bioproducts are moved about in public places between establishments and institutes. They can travel by hand, by public and private transport, by surface and air mail and by air freight. Passing through many hands during their journey, the containers may be involved in an accident and damaged or broken, releasing material; or may be lost or stolen and opened by unauthorized persons. It is important, therefore, to package such materials properly in accordance with national and international guidelines, and to have procedures in place for their safe handling [5,7,8]

References

1 Advisory Committee on Dangerous Pathogens. (1995) *Categorization of Biological Agents According to Hazard and Categories of Containment*, 4th Edn. HSE Books, Sudbury.

2. Frommer, W., Archer, L., Boon, B. *et al.* (1993) *Appl. Microbiol. Biotechnol.* **39**, 141.

3 Jones, B.P.C. and Morris, C.B. (1993) in *Cell and Tissue Culture: Laboratory Procedures* (A. Doyle, J. Griffiths and D. Newell, eds), 1A:2.1–1A:2.15. John Wiley & Sons, Chichester.

4 Richardson, J.H. and McKinney, R.W. (eds) (1993) *Biosafety in Microbiological and Biomedical Laboratories*. Publication no. (CDC)93-8395. US Department of Health and Human Services, Washington DC.

5. Health Services Advisory Committee. (1991) *Safe Working and Prevention of Infection in Clinical Laboratories*. HMSO, London.

6. Health and Safety Commission. (1997) *Control of Substances Hazardous to Health Regulations 1994. General COSH ACOP, Carcinogens ACOP, Biological Agents ACOP.* HSE Books, Sudbury.

7. Health and Safety Commission. (1996) *Carriage of Dangerous Goods (Classification, Packaging and Labelling) Regulations.* HSE Books, Sudbury.

8. International Air Transport Association. (1997) *Dangerous Goods Regulations*, 38th Edn. International Air Transport Association, Montreal.

Cold storage

Storage facilities require good housekeeping. They should be cleaned and checked periodically to remove unwanted and unlabeled items, broken containers and so on. Adequate records should be kept as to the contents of each of these storage facilities.

Training is needed in such operations as filling containers, transferring materials, venting storage vessels and dealing with condensates. The correct vessels must be used for liquefied gases. These vessels will be made from materials that are both compatible with the cryogens and are able to withstand the very wide and fairly sudden temperature variations that will be encountered [9].

Waste disposal

Waste disposal is one of the most critical and often the most neglected aspect of laboratory procedure. In the UK there is now a 'duty of care' imposed on all producers of waste, be they individuals or a corporate body [10]. This places even more responsibility on an establishment and its staff to ensure that waste materials are being disposed of in a safe and proper manner. When developing waste disposal policies the national legislative requirements and the requirements by national bodies will need to be considered. Each laboratory unit should develop procedures for the handling and storage of waste materials to ensure that they can be disposed of in a correct and consistent manner.

9. Simione, F.P. (1993) in *Cell and Tissue Culture: Laboratory Procedures* (A. Doyle, J. Griffiths and D. Newell, eds), 1C:1.1–1C:1.4. John Wiley & Sons, Chichester.
10 Department of the Environment. (1991) *Waste Management. The Duty of Care. A Code of Practice*. HMSO, London.
11. Ayliffe, G.A.J., Coates, D. and Hoffman, P.N. (1993) *Chemical Disinfection in Hospitals*. Public Health Laboratory Service, London.
12. Public Health Laboratory Service. (1993) *Safety Precautions, Notes for Guidance*. 4th Edn. Public Health Laboratory Service, London.
13. Cottam, A.N. (1989) *Specialist Inspector Report No. 17*. HSE Books, Sudbury.
14. British Standards Institution. (1995) *Safety Requirements for Electrical Equipment for Measurement, Control and Laboratory Use – Particular Requirements for Laboratory Centrifuges*. British Standards Institution, London.
15. Public Health Laboratory Service. (1993) *Health Hazard Note* [Hazard(93)8]. Public Health Laboratory Service, London.
16. Clark, R.P. (1983) *Occupational Hygiene Monograph No. 14* (reprinted 1989). Science Reviews, Leeds.
17. Stuart, D.D., Greenier, T.J., Rumery, R.A. and Eagleson, J.M. (1982) *Am. Ind. Hyg. Assoc. J.* **43**, 265–270.
18. British Standards Institution. (1992) *Microbiological Safety Cabinets*, BS5726, parts 1–4. British Standards Institution, London.

Disinfection

All waste materials must be made safe prior to disposal or removal from the laboratory [1]. Adequate decontamination procedures should be employed to ensure that such items are made safe. Disinfection, defined as the removal or destruction of pathogenic micro-organisms, is one such procedure. The use of chemicals for disinfection purposes does have some limitations and should only be used in situations where other, more reliable methods, cannot be employed. Disinfection should not be used when:

(i) sterilization is required;
(ii) physical methods can be employed;
(iii) thorough cleaning is adequate;
(iv) disposables can economically be used.

Therefore, in the tissue culture laboratory, the use of chemical disinfectants is limited to a few well-defined situations such as:

(i) the disinfection of instruments and apparatus when physical methods cannot be used;
(ii) making certain items safe for subsequent handling;
(iii) the decontamination of hard surfaces (e.g. bench surfaces) and 'housekeeping' some increased risk areas (e.g. spillages).

In the UK, the most commonly used disinfectants are clear phenolics and hypochlorites, with aldehydes having some limited application

19. Jones, B.P.C. (1993) in *Cell and Tissue Culture: Laboratory Procedures* (A. Doyle, J. Griffiths and D. Newell, eds), 1B:1.1–1B:1.7. John Wiley & Sons Ltd, Chichester.
20. Newsom, S.W.B. and Walsingham, B.M. (1974) *J. Clin. Pathol.* **27**, 921.
21. Health and Safety Executive. (1996) *Occupational Exposure Limits*. Guidance Notes EH40 (yearly). HSE Books. Sudbury.
22. Department of Health and Social Security. (1977) *Health Equipment Information* 70:56/77. Department of Health and Social Security, London.

Codes of laboratory practice

Code of laboratory practice and discipline
Waste disposal
Disinfection
Use of safety cabinets

Protocol provided

1. *Fumigation of safety cabinets*

[11]. In the USA, iodophors and quaternary ammonium compounds (QACs) are more popular [4]. Both hypochlorites and aldehydes are effective against viruses [11]. Hypochlorite solutions containing 20 000 ppm of available chlorine are effective against prions after 1 hour.

Problems: Avoid prolonged treatment of metals with hypochlorites; buffered hypochlorites (pH 7.5–7.6) are less corrosive than unbuffered. Hypochlorites should not be used on moving parts of machinery or equipment liable to stress (e.g. centrifuge rotors and tubes).

Use of centrifuges

The potential hazards from centrifuges are very considerable. They may arise from mechanical faults or from infection risks from the loads. The best defence against mechanical failure of any part of a centrifuge is to always purchase a model which conforms with the relevant national standards (e.g. ref. 14); to install it in the best available position [15] and at such a height that all staff can see into all parts of the bowl; to keep it clean, regularly inspected for signs of corrosion and properly serviced. The recommended procedures for balancing loads and operating the centrifuge must be followed for each particular model. The degree of risk in the centrifugation of biological agents will depend on a number of factors such as the type of apparatus used, the volume of material handled and the care with which it is manipulated. Aerosols may be created during dispensing or syphoning of suspensions and supernatant. A heavy aerosol may be created in the event of a bursting or overflowing of a tube, or of some mechanical fault. It is clear that unless the agent is confined, the procedure will be hazardous [1,3,5,12].

Problem: Disinfectants such as hypochlorites should not be used on metal centrifuge rotors and tubes or other equipment liable to stress (see 'Disinfection', p. 15).

Use of safety cabinets

Safety cabinets are designed to protect the laboratory worker from the inhalation of any aerosols that may arise during the work activity. The principle is that air is drawn over the working area and then exhausted to atmosphere through a HEPA filter. Such cabinets are not to be confused with laminar airflow cabinets (LAF), clean work-stations or fume cupboards as these should not be used for any work involving biological agents. All safety cabinets require careful siting, installation and regular maintenance [16,17] and are required to conform to the specifications laid down by the appropriate national regulatory bodies (e.g. ref. 18).

Problems: A majority of cabinets exhaust, via a HEPA filter, directly to the atmosphere. However, there will be some situations where this cannot be achieved. Cabinets which recirculate the air back into the laboratory are, therefore, used, but special permission is required [1]. Such cabinets must be fitted with two HEPA filters in series.

Fumigation of safety cabinets (see *Protocol 1*)

Disinfection using gaseous formaldehyde fumigation should be performed prior to their 6-monthly servicing and testing requirement [1,5]. On such occasions the cabinets can be fumigated at the same time as the laboratory, by leaving the front closures, night doors or front windows open. However, there may be occasions when the nature of the work (i.e. between experiments) or a spillage demands fumigation of the cabinet only. The generation of formaldehyde vapor from formalin, as detailed in *Protocol 1,* is the simplest way to carry this out [12,19,20].

Problems: A majority of cabinets exhaust direct to atmosphere, via a HEPA filter, and can be purged of formaldehyde simply by switching the cabinet on. Problems will arise, however, when the time comes to purge those cabinets that recirculate exhaust air back into the laboratory. However, this will require monitoring to ensure that the formaldehyde concentration levels are below the maximum exposure levels (MEL) of 2.5 mg/m^3 (2 p.p.m.; see ref. 21). Another alternative used in the USA for class II recirculating cabinets is to fit a length of flexible trunking over the exhaust and purge the formaldehyde to atmosphere via an open window or a fume hood.

Another problem is that formaldehyde can react with some chlorine-containing solutions to form a potent lung carcinogen [22].

Code of laboratory practice and discipline [3]. B. Jones

1 Ensure that approved protective clothing (i.e. laboratory coats, back-fastening gowns, gloves, safety spectacles, visors, aprons, etc.) is available and used, as indicated by laid-down procedures.

2 Always use techniques that avoid unecessary splashing and possible contamination of the working surfaces. For example, manipulate fluids slowly and gently using the appropriate pipetting aids. ①②

3 Wear the protective footwear available for specific activities (e.g. transport of gas cylinders, centrifuges and other heavy apparatus). Do not wear shoes that are unstable or do not protect the toes from falling objects.

4 Tie back long hair.

5 Do *not* eat, drink or smoke, apply cosmetics or perform any other action likely to contaminate body surfaces or street clothing.

6 Do *not* mouth-pipette under any circumstances. This is strictly forbidden.

7 Do *not* sit on apparatus, the bench or any other working surface.

8 Always use approved safety devices, such as safety cabinets and fume hoods, whenever they are indicated by procedures and assessments. ③

Notes

① Cells from any source may be infected with latent viruses or other adventitious agents, therefore uninoculated, as well as inoculated, cultures must be considered potentially hazardous. It is important to handle cell cultures so as to avoid contaminating the working environment, microbial contamination of the cultures and cross-contaminating different cell types.

② Procedures such as dissection, grinding and homogenization by manual, mechanical or ultrasonic methods should be carried out carefully, as these techniques generate splashes and aerosols which lead to contamination of the work area. Carry them out under the properly designed hoods and covers or, preferably, within a microbiological safety cabinet.

③ Ensure that such devices are properly serviced and maintained.

④ Keep the storage of laboratory consumables, packaging materials, etc., to a minimum.

⑤ For example, flammables in fireproof cupboards, glassware properly stacked on shelves.

9 Label all material clearly so that everybody can understand their nature, and use approved hazard notices, tape, etc., where appropriate.

10 Keep the laboratory area clean and free from clutter, and do not obstruct aisles and exits. ④ Do not hoard old cultures, reagents and so on, unless it is certain that they will be needed again.

11 Ensure that all equipment and materials are stored in a safe and proper manner. ⑤

12 Keep all unnecessary paperwork, books and writing implements away from work surfaces.

13 Ensure that all materials and apparatus are made safe by approved methods before leaving the laboratory area, and dispose of all unwanted materials in accordance with stated safety procedures.

14 Always report sickness and injuries to the appropriate authority.

15 Always report spills, accidents, or other incidents from which lessons may be learned, to the appropriate authority. ⑥

16 Remove laboratory clothing and wash hands before leaving the laboratory area.

⑥ Physical hazards must also be considered, such as needle-sticks from hypodermic syringes, needles and cannulae, cuts from scalpels and contaminated glassware such as capillary pipettes, microscope slides, etc. Suitable alternatives must be considered before the use of such items, such as the substitution of plastic to reduce glassware injuries. Every worker handling sharps should develop a safe routine for their use, transport and disposal.

Code of practice for waste disposal [5,12]. B. Jones

1 All waste materials must be made safe prior to disposal or removal from laboratories.

2 Segregate waste at source (i.e. within the laboratory). ①

3 Dispose of waste materials 'little and often' to avoid the problems arising from the bulk storage of such material.

4 Except for some specific situations all contaminated or potentially contaminated material from laboratories must be autoclaved prior to disposal or recycling. ②

5 Use adequate safe procedures and containers for the transportation of waste materials. ③

6 All sharps, whether used for infectious material or not, must be discarded in purposely designed 'sharps' containers and incinerated. ④

7 Autoclaves and incinerators must only be operated by trained authorized staff.

Notes

① This will avoid segregation by third parties who may or may not be aware of the nature of the waste and, hence, the subsequent hazards involved. Waste materials may be classified as solid or liquid or they can be categorized according to their nature and potential hazard [i.e. biological (infectious), chemical and radiological]. Other subdivisions may include disposables, recyclables, sharps and so on.

② Bottles and other vessels containing fluid should be autoclaved using the proper conditions and must not be tightly sealed. Cellulose nitrate centrifuge tubes and other such articles must not be autoclaved.

③ Autoclave bins must be used for the transfer of infected equipment and waste material to the autoclave.

④ If the sharps are contaminated with infectious material, these containers can be autoclaved but they must be finally disposed of by incineration.

1 Do not use disinfectants when other, more reliable, methods such as autoclaving or incineration can be used.

2 Use clear soluble phenolic disinfectants for most organic matter and general bacteriology, but not for blood and viruses.

3 Use hypochlorite-based disinfectants for minimal organic matter, small quantities of blood and viruses. It is recommended that the following concentrations of hypochlorite should be used in the laboratory:

- general use: solution containing 1000 p.p.m. of available chlorine;
- pipette and other discard jars: solution containing 2500 p.p.m. of available chlorine;
- blood spillages, etc.: solution containing 10 000 p.p.m. of available chlorine.

These solutions should be made up fresh each time (i.e. daily). Do not use on metals. ①

4 Keep a fresh supply of the appropriate disinfectants ready for use whenever infectious material is being handled. ②

5 Working surfaces should be disinfected regularly, at least at the end of

Notes

① Hypochlorites should not be used on moving parts of machinery or equipment liable to stress (e.g. metal centrifuge rotors and tubes). Glutaraldehyde (2%) is the disinfectant of choice for such equipment. As with hypochlorites, fresh solutions of glutaraldehyde are required. Care should be taken with its use, as it is pungent and can lead to irritation of the eyes and nose. If it is found to be unpleasant, then succinate dialdehyde is a suitable alternative.

② Some disinfectants deteriorate quickly after dilution, therefore, it is important that fresh stocks are made up at least weekly.

③ Most disinfectants, due to their nature, and especially in concentrated form, are corrosive and irritant to the skin and, more particularly, to the eyes. Great care should be exercised in handling the concentrate. Gloves should be worn and eye shields or face visors if splashing is likely. In some situations aprons or boots may also be required. In line with other chemical substances, COSHH assessments [6] need to be made on the use of these disinfectants, and the protective equipment required should be clearly stipulated.

the working day, as part of basic laboratory practice, especially those areas where blood and other body fluids are being handled.

6 Use aldehydes for special purposes only (e.g. see *Protocol 1*). ①

7 Wear disposable gloves when swabbing with disinfectants. ③

Code of practice for the use of safety cabinets [5]. B. Jones

1 Before use, check that the cabinet is working properly. ① Ensure that all materials and equipment are placed in the cabinet before the work is started. ②

2 Ensure that a vessel of the appropriate disinfectant (e.g. 70% ethanol) is on hand within the cabinet in case of spillages, etc.

3 Bear in mind that once work has started, all materials, etc., within the cabinet are potentially contaminated and should not be removed from the cabinet until after appropriate disinfection. ③

4 Work slowly and surely. Any fast movements will interrupt the airflows.

5 When the work has been completed, ensure that all materials and equipment are made safe. Place all materials that need to leave the cabinet in appropriate transport containers and disinfect either by spraying or wiping. Disinfect the working area in case of spillages and splashes (see 'Code of practice for disinfection').

6 Remove containers from cabinet. Transport safely to incubators, cold rooms, autoclave bins and so on.

7 Leave the cabinet running for 5–10 min before switching off. ④

Notes

① This will depend on the class of cabinet but, generally speaking, check the airflow indicators or negative pressure gauges. Many cabinets are now fitted with alarms to indicate an 'unsafe' condition.

② The less traffic in and out of the cabinet, the better, as this interrupts the airflows and will reduce the risk of contamination.

③ This will include gloved hands.

④ Depending upon the work being carried out, the cabinet may need to be fumigated with formaldehyde prior to further work being undertaken (see *Protocol 1*).

⑤ The use of flammable gases within safety cabinets should be strictly forbidden.

Protocol 1. **Fumigation of safety cabinets.** B. Jones

Reagents

Formalin (40% formaldehyde) ⚠
Potassium permanganate ($KMnO_4$) (if necessary, see step 2) ⚠

Equipment

Formaldehyde monitoring equipment (if necessary, see note 3)
Temporary exhaust ducting (if necessary, see note 3)
Thermostatically controlled heater or 500 cm^3 beaker

Procedure [19]

1 Place 25 ml commercial formalin (40% formaldehyde) in a purpose-built thermostatically controlled heater inside the cabinet. ① Close the window or replace the front closure/night door.

2 Should a heating device be unavailable, place 35 ml commercial formalin in a 500 ml beaker inside the cabinet. Add 10 g potassium permanganate and seal the cabinet. ②

3 Leave for at least 6 h, preferably overnight.

4 Switch on the cabinet fan and remove the front closure/night door (if fitted), to allow the formaldehyde to exhaust to atmosphere. ③

5 Inform the necessary authority of the intention to discharge formaldehyde via the cabinet exhaust so that any maintenance engineers are forewarned. ④

Notes

This procedure takes approximately 18–24 hours.

① The actual quantity of formalin used will depend upon the size of the cabinet. Allow 2 ml formalin for every 0.028 m^3 of space. The volume of formalin quoted is that for a cabinet having an external air volume of 0.38 m^3 [20].

② The mixture will boil exothermically in a few seconds, releasing formaldehyde. Excess potassium permanganate may create an explosive mixture.

③ Care should be taken while fumigating cabinets with double HEPA filters that recirculate air within a laboratory (see 'Methods available').

④ Avoid excessive or unnecessary fumigation of safety cabinets as it may adversely affect the running of the cabinet and result in a deposition of formaldehyde polymers on cabinet surfaces.

II BASIC INGREDIENTS FOR MEDIA. C. Morris

Media formulations: practical aspects, preparation and storage

In vitro growth of cell lines requires a sterile environment in which all of the nutrients for cellular metabolism can be provided in a readily accessible form at the optimal pH and temperature for growth. A large number of defined media are commercially available which are suitable for supporting the growth of the majority of cell lines. Their formulations have been designed by empirically measuring the optimal concentration of each ingredient for a specific cell line or cell type [1–3].

Defined media contain inorganic salts, which provide buffering capacity and osmotic balance to counteract the effects of carbon dioxide and lactic acid produced during cellular metabolism. Other components are amino acids, vitamins and cofactors. Energy is usually supplied in the form of glucose or galactose. Formulae vary in complexity and have been developed to support a wide variety of cell types, examples of which are:

- Medium 199 was originally formulated to support the growth of chick embryo fibroblasts and later mouse L cells.
- Eagles' basal medium is suitable for primary mammalian cells and human diploid cells.
- Dulbecco's modified Eagles' medium (DMEM) supports the growth of a wide range of mammalian cell lines, including hybridomas.
- RPMI 1640 is an enriched medium suitable for human leukemia cells, but also supporting a wide range of mammalian cells, including hybridomas.
- Glasgow minimum essential medium was specially formulated for BHK-21 cells.
- Ham's F12 was originally formulated for cloning Chinese hamster ovary cells.
- Iscove's modified Dulbecco's medium is an enriched medium suitable for supporting high-density growth.
- Liebovitz's L-15 medium is designed for use in carbon dioxide-free environments.
- TC 100 insect medium was developed to support the growth of insect cell lines (e.g. *Spodoptera frugiperda*).

It is essential that the water used is as free from all mineral, organic and microbial contaminants as possible when media are to be prepared from basic ingredients or commercially prepared mixtures. Ideally this should be from a pyrogen-free, reverse osmosis- and resin cartridge-purified source, with a final resistance of 16–18 MΩ. All filtration equipment and storage bottles must be free from trace contamination of detergents and preferably sterilized by autoclaving or dry heat. Although it is possible to prepare media from basic ingredients, this is time-consuming and often no less expensive than using commercially prepared ready-mixed powders.

As the finished medium must be sterile, it should be prepared in a room separate from other activities and by staff wearing clean or sterile masks, gowns, caps and gloves during all stages of preparation. Prepared media must be tested for sterility and their ability to support the growth of cells.

Serum screening and selection

Serum is a complex solution of albumins, globulins, growth promoters and growth inhibitors. The majority of cell lines require the addition of serum to defined culture medium to stimulate growth and cell division. The quantity and quality of these components will be affected by the age, nutrition and health of the animals from which the serum is obtained (i.e. it is subject to significant biological variation). Cell lines require very high quality serum to promote optimum growth and so fetal or newborn bovine serum is used more widely than serum obtained from other sources. The following points should be taken into account when purchasing animal serum:

(i) choose a supplier who either processes their own serum or obtains it from a veterinary-approved collection point. A reputable supplier will be able to provide all of the documentation necessary to trace the serum from source to bottling (i.e. a Certificate of Origin).
(ii) If cells are to be exported (e.g. to the USA) restrictions may be imposed on cell lines grown in serum obtained from countries with indigenous livestock carrying transmissible agents (e.g. bluetongue, foot-and-mouth disease, infectious

bovine rhinotracheitis). Obtain a Certificate of Analysis. This lists the quality control tests undertaken on the final product including the results of virus and *Mycoplasma* screening. Some points to look for are:

- infection of the serum by gram-negative bacteria during collection and processing will be indicated by the presence of endotoxin. The highest quality sera will have levels below 0.2 ng/ml in fetal bovine serum;
- levels of hemoglobin released from erythrocytes during initial processing should be low (i.e. less than 20 mg/100 ml in fetal bovine serum).

Methods available

Preparation of media from basic ingredients (see *Protocol 2*)
This will require a complete medium formula of all the ingredients and their weights (e.g. DMEM). If any of the organic salts are hydrated, the correct form must be used to obtain the correct weight. Amino acids may be derived from microbial sources and should be handled cautiously as they will be a potential source of contamination. Other components (e.g. selenium, Phenol red) may be toxic and need to be handled with suitable containment and, in the UK, observing the relevant COSHH requirements. The advantage of preparing media this way is that all the ingredients are fresh and the finished product should give maximal growth support.

Preparation of media from premixed powders
Before preparing a batch of medium, check the expiry date on the

References

1. Morgan, J.G., Morton, H.J. and Parker, R.C. (1950) *Proc. Soc. Exp. Med.* **73**, 1.
2. Eagle, H. (1959) *Science* **130,** 432.
3. Ham, R.G. and McKeehan, W.L. (1978) *In Vitro* **14**, 11.
4. Sato, G.H., Pardee, A.B. and Sirbasku, D.A. (1982) *Cold Spring Harbor Conference on Cell Proliferation,* Vol. 9, *Growth of Cells in Hormonally Defined Media.* Cold Spring Harbor Laboratory Press, Cold Spring Harbor, NY.

Protocols provided

2. *Preparation of cell culture medium from basic ingredients*
3. *Plating efficiency of adherent cell lines*
4. *Cloning efficiency of suspension cell lines*
5. *Growth promotion*

container. Powdered media generally have a shelf-life of 24–36 months. Components vary in their stability, and may need replacing on preparation (e.g. vitamins; L-glutamine).

Serum preparation

A baseline (i.e. control serum) is needed for comparison with the serum batches to be tested. Normally the current batch in use is taken but if one is not available, select several batches, and use the one giving the best results as the control. If a particular batch is superior to those currently or previously tested, store aliquots at −20°C or, preferably, −70°C for future tests.

The tests undertaken measure the ability of the serum to support three parameters: cell attachment and spreading (plating efficiency); cloning efficiency; and growth promotion (cell division). All tests should be conducted in antibiotic-free medium with 10% serum (v/v) unless a different serum concentration is normally used.

Plating efficiency (see *Protocol 3*)

This measures the efficiency of cell attachment and spreading at low cell densities. Cell lines suitable for this test are the human lung carcinoma A549 and the Chinese hamster ovary CHO-K1.

Cloning efficiency (see *Protocol 4*)

This is appropriate for suspension lines where it is important to know the survival of cells at very low densities [e.g. myelomas (P3X63ag8.653 and Sp2/0) and hybridomas]. Cells are diluted to a range of densities around one cell per unit volume of medium.

Growth promotion (see *Protocol 5*)
In this test the ability of serum to provide growth factors is measured by estimating the number of population doublings within a given time. From this it is possible to determine the cell doubling time. The human fetal diploid line MRC-5 is particularly suitable for this test, as its growth is especially influenced by the quality of the serum.

Protocol 2. **Preparation of cell culture medium from basic ingredients.**

C. Morris

Reagents

200 mM L-Glutamine
1 M HCl
1 M NaOH
Nutrient broth (Tryptone soya broth or thioglycollate)
Phenol red ⚠
Pre-weighed formula components
Pure pyrogen-free water

Equipment

37°C Incubator
Magnetic stirrer
Mixing vessel with a Teflon-coated magnetic stirring bar ①
pH meter
Sterile bottles
0.2 μm Sterile filtration apparatus and pressure vessel
Weighing scales

Procedure

1 Pre-weigh all of the components in about 90% of the final volume of water at room temperature. In the case of any mineral salts check that the salt formula matches the molecular formula given in the selected media in terms of valency and hydration.

2 Mix with a magnetic stirrer until completely dissolved and then add additional requirements (e.g. sodium bicarbonate, L-glutamine and Phenol red as a pH indicator).

Notes

This procedure takes approximately 3 hours.

① Use a mixing vessel that is at least 25% larger in volume than the final medium volume.

② As filtration usually raises the pH by 0.1–0.3 units, an allowance should be made to compensate for the rise (e.g. adjust to pH 7.0 for a post-filtration pH of 7.2).

③ When flat-bed filters are used it is advisable to use a filter stack consisting of a pre-filter, a 0.45 μm filter and a final

3 Test the pH and adjust to the desired pH using either 1 M HCl or 1 M NaOH. ②

4 Add water to bring the medium to the required volume and sterilize by filtration through a sterile filter of 0.2 μm or less. ③

5 The final volume of prepared medium should allow for the addition of serum and other supplements (e.g. tryptose phosphate broth). For example a 10 liter batch should be made up to a final volume of 9 liters if 10% serum is to be added, and the 900 ml aliquots dispensed into numbered, sterile 1 liter bottles. Store at 4°C in the dark. ④

6 Microbial quality control checks should be made on at least the first and last bottle collected after filtration, but preferably on each bottle (i.e. 2–5% of the total bottle volume is removed aseptically for testing). Add the medium to a broth suitable for general incubation of micro-organisms in a ratio of 1:3 (e.g. Tryptone soya broth or thioglycollate) and incubate at 37°C for 5–7 days. Reject any bottles that are positive for microbial growth rather than refiltering them. ⑤

0.2 μm filter at the base.

④ Only sufficient medium for 4–8 weeks should be prepared at one time, to ensure maximum cell growth. When medium is kept for longer than 4 weeks, fresh glutamine (i.e. from a 200 mM stock solution) should be added (1 ml/100 ml medium).

⑤ Sodium bicarbonate is omitted from media to be used in CO_2-free environments (e.g. Leibovitz L15).

Protocol 3. **Plating efficiency of adherent cell lines.** C. Morris

Reagents

Carnoy's fixative (1 part glacial acetic acid:3 parts methanol) ⚠
Cell culture medium containing 10% test or control serum
Cell cultures: A549 and/or CHO-K1
1% (v/v) Crystal violet
0.4% (w/v) Trypan blue in phosphate-buffered saline ① ⚠
Trypsin/ethylenediamine tetraacetic acid (EDTA)

Equipment

Bench-top centrifuge with swing-out rotor
5–10% CO_2 incubator at 37°C
15 ml Conical screw-capped tubes
Hemocytometer
Inverted microscope
Pipettes plus pipettor
Six-well cell culture plates with lids

Procedure

1 The cell line(s) to be used must be healthy and in the exponential growth phase (i.e. actively dividing). Harvest the cells in the same way as for routine sub-culture, for example using trypsin/EDTA (*Protocol 8*), and distribute them equally into the same number of 15 ml conical screw-capped tubes as the serum samples to be tested. Pellet the cells by centrifugation at 70–100 *g* for 5 min at room temperature. Decant the supernatants and resuspend each pellet in a small volume of medium containing either the test serum(s) or the control serum (i.e. one pellet for each serum).

2 Perform a viable cell count for each tube using a 1:4–1:10 dilution of cells to Trypan blue. For each serum prepare 12 ml of medium

Notes

This procedure takes approximately 3 hours to set up, up to 15 days to complete.

① Trypan blue is harmful if ingested or inhaled. It is irritating to the eyes, harmful on skin contact and has been found to cause cancer in laboratory animals. Appropriate precautions should be taken when handling Trypan blue and the use of an extraction hood and gloves is advised.

② Sera with relative PEs of less than 75% (i.e. against control serum) should not be selected if plating efficiency is a key criterion for selection. When calculating the results it is important to compare the same number of wells for each serum tested.

containing 6×10^2 cells and inoculate the six-well plates (3.5 cm diameter wells) with 1×10^2 viable cells/well (i.e. 2 ml of medium).② Alternatively use six individual 3.5 cm culture dishes. Place the plates in a CO_2 incubator at 37°C.③

3 At 3, 4 and 5 days post-inoculation examine for the growth of colonies at ×100 magnification on an inverted microscope. Continue to examine the colonies until they are of a size large enough to be seen by the naked eye, but are not merging with each other. Fix the cells by adding 2 ml Carnoy's fixative directly to the medium and standing for 10 min at room temperature. Remove and immediately add a further 2 ml of fixative. Stand for 5 min and decant, allowing the cells to air-dry. Plates are usually examined, fixed and stained 5–15 days post-inoculation.

4 Stain with Crystal violet (1% v/v) for 10 min at room temperature. Wash the cells in distilled water and leave the plates to dry. Count the number of colonies in each well and record, then calculate the plating efficiency. ④

Table 2. Calculation of plating efficiency

	Colonies/well						Total	PE
Test serum	62	73	66	60	69	76	= 406	67.7 %
Control serum	78	68	71	75	77	65	= 434	72.3 %

③ It may be necessary to increase the CO_2 level in the incubator to compensate for the slow growth of the cells. Note the color of the medium, which should be red to orange. Purple medium indicates that the pH may be too high (i.e. >pH 8).

④ Plating efficiency (PE) is calculated by dividing the total number of colonies in each set of wells by the number of cells inoculated (i.e. $6 \times 100 = 600$). An example of results obtained would be as shown in *Table 2*.

Relative PE for test serum vs. control serum =

$$\frac{\text{PE \% of test serum (67.7)}}{\text{PE \% of control serum (72.3)}} \times 100 = 93.6\%$$

Protocol 4. **Cloning efficiency of suspension cell lines.** C. Morris

Reagents

Cell culture: myeloma or hybridoma
Cell culture medium containing 10% test or control serum
0.4% (w/v) Trypan blue in phosphate-buffered saline⚠

Equipment

Bench-top centrifuge with swing-out rotor
96-well Cell culture plates with lids
10% CO_2 incubator at 37°C
Hemocytometer
Microscope
Pipette tips plus pipettor
Sealing tape

Procedure

1 Perform a viable cell count on the cell culture using Trypan blue (*Protocols 13* and *14*). The cells should only be used if the viability is more than 90%.

2 Inoculate 96-well flat-bottomed plates with a series of dilutions to give 3, 1, 0.3 and 0.1 cells/well in 0.2 ml medium using a minimum of 16 wells/dilution. The cells are diluted from the control culture in the test media after making a viable count using Trypan blue exclusion.①

3 Place the lids on the plates and seal with tape to prevent evaporation if

Notes

This procedure takes approximately 2 hours.

① If the quality of the sera being tested is poor it may be necessary to extend the dilution range to five cells/well to provide sufficient data.

② When selecting a serum, an average CE of more than 80% is recommended. It is important to monitor the pH and possible evaporation within the plates as the cells will be very sensitive to small fluctuations during the first few days. Check the CO_2 and humidity levels of the incubator prior to commencing and thereafter daily.

they are not designed to reduce vapor loss. Incubate at 37°C in a 10% CO_2 incubator. Examine after 5 days then daily for up to a further 5 days if colonies are not well formed.②③

4 Count the number of wells with cell growth for each dilution. The cloning efficiency (CE) for each dilution of the test serum(s) is calculated by dividing the number of wells with growth by the number of wells for the control for the same dilution.

An example of results obtained would be as shown in *Table 3*.

Table 3. Calculation of cloning efficiency

	No. of wells with growth (out of 16)				
	3	1	0.3	0.1	= Total
Test	16	10	3	0	29
Control	16	12	5	0	33
Cloning efficiency	100	83	60	0	88

Compare the CE of the test and control for each dilution. In addition, calculate the average cloning efficiency by dividing the total number of test wells by the total number of control wells.

③ The addition of hypoxanthine and thymidine may be necessary to support the growth of some myelomas and hybridomas.

Protocol 5. **Growth promotion.** C. Morris

Reagents

Cell culture: human diploid fibroblast (e.g. MRC-5, WI-38)
Cell culture medium containing 10% test or control serum
0.4% (w/v) Trypan blue in phosphate-buffered saline⚠
Trypsin/ethylenediamine tetraacetic acid (EDTA)

Equipment

Bench-top centrifuge with swing-out rotor
25 cm^2 Cell culture flasks
5–7% CO_2 incubator at 37°C
Hemocytometer
Microscope
Pipette tips plus pipettor

Procedure

1 The cell line(s) to be used must be healthy and in the exponential growth phase (i.e. actively dividing).① Harvest the cells in the same way as for routine sub-culture (e.g. using trypsin/EDTA; *Protocol 8*). Inoculate 25 cm^2 flasks each with 5×10^5 cells in the test and control media (6–7 ml per flask). Prepare duplicate flasks.

2 Subculture the flasks every 4 days, perform viability and total cell counts and inoculate new flasks with 5×10^5 viable cells in the appropriate medium. Repeat this procedure for three further subcultures, recording cell counts each time.②

Notes

This procedure takes approximately 4 hours.

① A cell's optimum growth rate depends on several criteria (e.g. temperature, pH, serum concentration). Unlike attachment factors, the level of growth factors has a proportional effect on growth rate. Sera should therefore be selected for maximum growth, with a suggested minimum acceptance of 80% of the control cell-doubling time.

② When comparing several sera, any one particular batch may not give the best results for all the tests. If this

3 Estimate the number of population doublings using the equation:

$\log_{10}$ (harvest cell count) − $\log_{10}$ (inoculation cell count)/0.301.

Using the number of population doublings, estimate the average cell-doubling time (tD) by,

$$tD = \frac{\log_n}{u} (= 0.693)$$

where u = specific growth rate = $\log_n (X/X_o)/t$, $\log_n$ = natural logarithm, X = final cell density, X_o = initial cell density and t = time (h).

Compare the cell-doubling times in the test sera against the control serum. In general it can be concluded that the shorter the doubling time the better the serum. Results can, if necessary, be expressed as a percentage of the control serum.

happens it may be necessary to repeat the tests at a lower serum concentration in order to increase the test's sensitivity (e.g. between 3% and 5%).

III PRIMARY CULTURES. L. Haynes, V. Worthington, C. Morris and D. Newell

Explant cultures. L. Haynes

The tissue culture method was first employed by the developmental biologist Ross Harrison [1] in the study of growth in explant cultures of the amphibian central nervous system. It was developed further by Carrel [2] and later by others. The technique has remained the preferred experimental tool to the present day amongst a minority of tissue culture workers, but it has never grown popular because of its labor-intensiveness and the level of experience required to optimize growth conditions. However, it proves a very powerful and revealing model of normal development and pathological progression in the hands of an expert when coupled with appropriate equipment for anatomical and physiological investigation. Collectively, explant cultures from fetal or adult normal and pathological tissues from man and animals provide optimal cytodifferentiation and tissue–cell interactions, and in many cases provide insights into tissue structure development and pathological progression which are not possible *in situ*. The principal disadvantage of explant culture procedures is the high rate of culture failure.

A clean suitably equipped area should be set aside for operating work in the preparation room. The operator should wear clean disposable clothing and a face-mask for close work. It is undesirable to carry out whole-animal dissection in a laminar airflow cabinet. If necessary, a simple cabinet with a large operator access area may be dedicated for operations. Animals should never be operated on in the same cabinet as is used for culture preparations.

Culture vessels

The successful cultivation of explants in all types of format is critically dependent on the quality and cleanliness of the culture vessels. Very few vessel types are supplied commercially as pre-sterilized disposable products in culture-grade plastics, the only alternative vessels are normally made of re-usable glassware and must be rigorously cleaned, sterilized and recycled under the operator's direction. The principal items for culture chamber/vessel assembly are listed below and are of borosilicate glass unless otherwise indicated.

- Organ culture chamber,
- Maximow depression slide,
- re-usable Cruickshank chamber (Perspex),
- coverslips (no. 1–2 thickness, 16–22 mm round or square),
- De Long rotation culture flask.

Some of these items are supplied 'precleaned' (e.g. coverslips) but cleanliness to culture quality can never be assumed and steps must be taken to both clean and prepare glassware for use. It is therefore worthwhile to prepare large batches of glassware at one time. Two alternative methods for glassware cleaning are given in *Table 4*. For reaggregation cultures, glassware should be siliconized with dimethyldichlorosilane (e.g. from Sigma Chemical Co.) vapor and then washed with deionized water before sterilizing.

Anchorage of adherent explants

The substratum on to which an explant culture is attached must provide both a suitable biological surface and sufficient mechanical strength to retain it when the culture chamber is moved. Reconstituted collagen matrices are commonly used. The source of collagen for this purpose should be animal tendon (e.g. rat tail, bovine achilles; type 1). Collagen type 1 is best prepared freshly in the laboratory from rat tail. Growth of explants in a plasma clot is also possible, and has the advantage that the tissue derives a source of nutrition from the clot itself. Furthermore, the formation of the clot is more readily controlled by inserting the tissue explant in a small drop of gelling agent. Thrombin (0.1 U/ml) or chick embryo extract (50%) both activate the enzymic crosslinking of fibrin if chicken plasma is used.

Table 4. Alternative methods for cleaning glassware

Protocol A		Protocol B[a]	
Wash	Time	Wash	Time
Contra D-70 (5% v/v)	12 h	2% Isoclean 100°C in water	5 min × 2
Deionized water	3 × 10 min	Hot tapwater	2 h running
Dry		1% HCl	12 h
Sterilize		Cold tapwater	6 h running
		Deionized water	3 × 1 h
		Acetone	2 × 15 min
		Dry	
		Sterilize	

[a] New glassware requires an additional cleaning step in fuming nitric acid or concentrated sulfuric acid : No-chromix oxidant 1:1.

Explant culture formats

Primary organized cultures are prepared from whole intact tissue explants set up as organ cultures or organotypic (histiotypic) cultures, or they are reconstructed from two or more components, of which at least one may be a tissue explant. This chapter describes the methods for obtaining, transferring and maintaining suitable tissue explants. In all of the related culture methods, tissues are microdissected and transferred in an orientation-selected manner to the chosen culture chamber. Organ cultures consist of explants or multi-explant tissue constructs grown at the medium/gas interface. Organotypic cultures consist of explants grown submerged in medium for at least a part of the culture duration. Growth is usually in medium supplemented with a high percentage (15–40%) of serum, but for some culture types, serum-free media compositions may be possible or even desirable.

Enzymic techniques for tissue dissociation. V. Worthington, C. Morris and D. Newell

The procedures used to isolate cells from tissue must be designed to maximize the yield of functionally viable and dissociated cells remaining after digestion. Cultures can be initiated from a variety of tissues (i.e. embryonic, normal and tumor). When a choice is possible, embryonic tissue is often the simplest source from which cell cultures can be derived. Despite the widespread use of enzymes for tissue culture, the mechanisms of action are poorly understood.

It may be necessary to perform some mechanical separation prior to the use of enzymes. This can be achieved by first cutting up tissues finely using scalpels. Ideally, pieces of tissue should be 1 mm^3 or smaller, to maximize the surface area available for enzyme digestion. Tissues that are to be dissected should be placed in serum-free medium to provide an isotonic bathing solution. When cell cultures are produced from tissues of the respiratory, digestive and genito-urinary systems, the addition of antibiotics is advisable to prevent infection during the preliminary culture stages. Reconstituted enzymes should be aliquoted and stored at −20°C; repeated freeze–thaw cycles should be avoided.

Methods available

Preparation of tissue explants (*Protocols 6* and *7*)
All human tissue samples must be treated as a potential source of infectious organisms and the appropriate precautions must always be taken. For general guidance, *Table 5* shows the main primary organized culture formats. The three groups of explant cultures are organ, organotypic and reconstructed cultures [1]. The protocols given below describe only the initial stages of tissue preparation and transfer *in vitro*.
Problems: Overgrowth of cultures by mesenchymal cells/fibroblasts cannot be controlled. Good cultures can be achieved only by very

References

1. Harrison, R.G. (1907) *Anat. Rec.* **1**, 116.
2. Carrel, A. (1912) *J. Exp. Med.* **15**, 516.
3. Silbermann, M. and Manor, G. (1984) in *Methods of Calcified Tissues Preparation* (G.R. Dickson, ed.), pp. 467–530. Elsevier/North Holland, Amsterdam.
4. Costero, I. and Pomerat, C.M. (1951) *Am. J. Anat.* **89,** 405.
5. Toran-Allerand, C.D. (1990) in *Methods in Neurosciences,* Vol. 2, *Cell Culture* (P.M. Conn, ed.), pp. 275–296. Academic Press, New York.
6. Flynn, D., Yang, J. and Nandi, S. (1982) *Differentiation* **22,** 191.

careful dissection of extraneous contaminating connective tissue and by repetition. Necrosis cannot be avoided because of the limitations of gas exchange, particularly in static cultures, and the ubiquity of phagocytes. It can be minimized by keeping explant size to a minimum, paring explants to optimize surface area/volume ratio and avoiding tissue damage during dissection.

Enzymic dissociation

Generally all of the enzymes, except trypsin, can be directly dissolved in a balanced salt solution or buffer of choice [19]. A stock solution of trypsin should be made initially by reconstituting the enzyme in 0.001 M HCl. This solution can be diluted into the digestion balanced salt solution or buffer immediately prior to use.

Collagenase: A brief description of cell dissociation enzyme and the target tissues for which it is mainly used is given below. Four types of mammalian collagenase are identified and all are available commercially:

(i) *type 1* contains average amounts of assayed activities (collagenase, caseinase, clostripain and tryptic activities). It is generally recommended for fat, adrenal and liver cells;

(ii) *type 2* contains greater clostripain activity. It is generally used for heart, bone, muscle, thyroid, cartilage and liver;

(iii) *type 3* is selected for low proteolytic activity. It is generally used for mammary cells;

7. Regnier, M., Prunieras, M. and Woodley, D. (1981) *Front. Matrix Biol.* **9**, 4.
8. Solursh, M. (1986) in *Articular Cartilage Biochemistry* (K. Keuttner, R. Schleyerbach and V.C. Hascall, eds), pp. 145–161. Raven Press, New York.
9. Marcel, M., Kint, J. and Meyvisch, C. (1979) *Virch. Arch. B. Cell. Pathol.* **30**, 95.
10. Honegger, P., Lenoir, D. and Favrod, P. (1979) *Nature* **282**, 305.
11. Ringer, S. (1895) *J. Physiol. (Lond.)* **18**, 425.
12. Tyrode, M.V. (1910) *Arch. Int. Pharmacodyn. Ther.* **20**, 205.
13. Parker, R.C. (1961) *Methods of Tissue Culture,* 3rd Edn, p. 57. Harper, New York.
14. Gey, G.O. and Gey, M.K (1936) *Am. J. Cancer* **27**, 55.
15. Earle, W.R. (1943) *J. Natl Cancer Inst.* **4**, 165.
16. Puck, T.T., Cieciuira, S.J. and Robinson, A. (1958) *J. Exp. Med.* **108**, 945.
17. Hanks, J.H. and Wallace, R.E. (1949) *Proc. Soc. Exp. Biol. Med.* **71**, 196.
18. Dulbecco, R. and Vogt, M. (1954) *J. Exp. Med.* **99**, 167.
19. Freshney, R.I. (1982) *Culture of Animal Cells: a Manual of Basic Techniques.* Alan R. Liss, New York.

Protocols provided

6. *Preparation of human tissue explants*
7. *Preparation of animal tissue explants*

Table 6. Culture formats

Culture format	System	Duration	Principal application[a]	Pilot ref.
Organ	Organ culture chamber	3 days to 2 months	d,g,h,t,u,v	3
Organotypic	Rocker chamber/roller tube	2–9 months	c,d,h,p	4,5
	Maximow slide assembly/ Cruikshank chamber			
	Gelfoam/collagen matrix			6
	Dermal equivalent			7
Reconstructed tissue cultures	Dermal equivalent	1–2 weeks	c,d,g,h,u	8
	Microwell/cell dot[b]			
	Confrontation			9
	Reaggregating[b]			10

[a]Applications code: c, cell interaction; d, differentiation; g, growth kinetics; h, histopathology; m, microanatomy/cytology; p, physiology; t, tumorigenesis; u, ultrastructure; v, virology.
[b]These systems are prepared from dissociated tissue cells, although they take on the appearance of organized three-dimensional cultures.

(iv) *type 4* is selected because of low tryptic activity. It is commonly used for islets and other applications where receptor integrity is crucial.

Trypsin: Trypsin is a pancreatic serine protease. It is one of the most highly specific proteases known, although it also exhibits some esterase and amidase activity.

'Trypsin' is the name commercial suppliers have given to pancreatin, a crude mixture of proteases, polysaccharidases, nucleases and lipases extracted from porcine pancreas. Crude 'trypsins' (like NF 1:250 and 1:300) are widely used for dissociating tissues. It appears, however, that crude trypsin and crude collagenase dissociate tissues by different mechanisms; difficulties are often encountered when using NF 1:250 preparations, the most common being incomplete solubility, batch to batch variability and cell toxicity.

In tissue culture laboratories, researchers use purified trypsin to release cells into suspension from monolayers growing on the surfaces of culture vessels. All proteolytic enzymes can undergo autolysis. They should therefore be dissolved immediately before use.

Elastase: Pancreatic elastase is a serine protease which also exhibits esterase and amidase activity. While elastase will hydrolyze a wide variety of protein substrates, it is unique among proteases in its ability to hydrolyze native elastin, a substrate not attacked by trypsin, chymotrypsin or pepsin.

Elastase is frequently used to dissociate tissues which contain extensive intercellular fiber networks. It is generally used for this purpose with other enzymes such as collagenase, trypsin and chymotrypsin.

Hyaluronidase: Hyaluronidase is a polysaccharidase with a specificity for bonds common in hyaluronic acid and chondroitin sulfate A and C. Because these substances are found in high concentrations in virtually all connective tissues, hyaluronidase is often used for the dissociation of tissues, usually in combination with a crude protease such as collagenase

Papain: Papain is a sulfhydryl protease. It has wide specificity and will degrade most protein substrates more extensively than the pancreatic proteases but is less damaging to cells [11].

Deoxyribonuclease I: Nucleic acid, especially DNA, often leaks into the dissociation medium as a result of cell damage, increasing the viscosity of the medium, which causes handling problems. Deoxyribonuclease will digest the fibrous strands without damaging intact cells.

Pronase: This protease is a bacterial enzyme. It is often used as an alternative to trypsin for subculturing cell lines. Because of its mild action, pronase may have to be used in conjunction with another enzyme (e.g. collagenase).

Ethylenediamine tetraacetic acid (EDTA): EDTA at a concentration of 0.02% (w/v) in calcium- and magnesium-free phosphate-buffered saline, can be used in conjunction with trypsin to enhance tissue disaggregation. Acting as a chelating agent, it dissociates intercellular links such as tight junctions. For this reason it is not suitable for use with dispase.

Choice of enzymic technique: There is a complex relationship between cell yield and viability. In general, there is an area of optimized recovery balanced between yield and viability. The relative digestive effectivenesses of the commonly used enzymes are given in *Figure 2*.

Improving technique:

(i) Low yield/low viability. Over/under dissociation, cellular damage. Change to a less-digestive-type enzyme and/or decrease the working concentration (e.g. from trypsin to collagenase, from type 2 collagenase to type 1).

(ii) Low yield/high viability. Under-dissociation. Increase the enzyme concentration and/or incubation time and monitor both yield and viability response. If the yield remains poor, evaluate a more highly digestive-type enzyme and/or the addition of a secondary enzyme(s).

(iii) High yield/low viability. Good dissociation, cellular damage. Enzyme overly digestive and/or at too high a working concentration. Reduce the concentration and/or incubation time, and monitor yield and viability response. Try diluting the proteolytic action by adding bovine serum albumin (0. 1–0.5%, w/v). Try

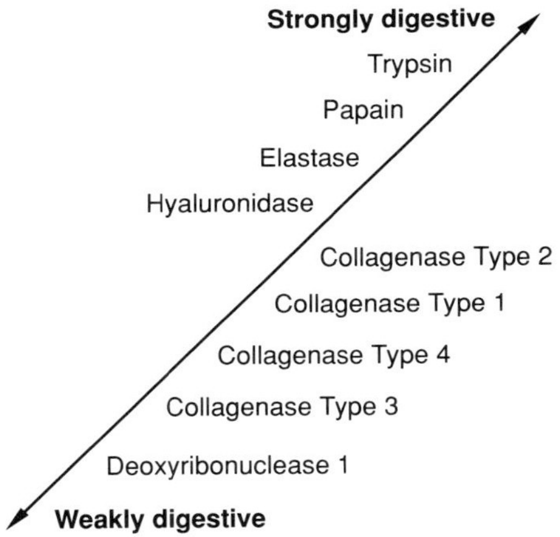

Figure 2. Enzyme digestion scale.

using a less proteolytic enzyme, although yield may be affected and should be monitored.

(iv) High yield/high viability. The ideal situation. Consider evaluating the effect of dissociation parameters to learn their limitations for future reference.

Protocol 6. **Preparation of human tissue explants.** L. Haynes

Reagents

Eagle's minimal essential medium (MEM) with penicillin and streptomycin (500 U/ml each), without serum
Hank's balanced salt solution (HBSS) without Phenol red
10% Hibitane or 70% ethanol⚠
Human tissue samples ①⚠

Equipment

Disposable operator face-masks
Four or five sterile petri dishes
Iridectomy scissors
Medium scissors (optional)
Operating microscope (×1–3 objectives)
Presterilized latex surgical gloves
Spemann pipettes with 0.6–1.0 mm orifice
Sterile distilled water
Sterile paper tissues
Sterilized stainless steel dissecting instruments
Tightly capped specimen jar
Two electrolytically sharpened mounted needles (optional)
Two microscalpels or two small scalpel handles with presterilized no. 11 or 15 blades
Two pairs watchmaker's forceps, straight and curved

Procedure

1 Wash the tissue pieces free of blood clots, plasma, unwanted biological tissues and visibly necrotic material.②

2 Cut the large (>0.5 cm) tissue fragments into smaller (2 mm cube) fragments in HBSS (10 ml in a petri dish).

3 Transfer the fragments twice to fresh HBSS in petri dishes.

Notes

This procedure takes approximately 1 hour after delivery of tissue samples.

① It is essential to establish and maintain a positive working relationship with the surgeon, midwife and, where appropriate, the histopathologists. It is essential that you are aware of any possible infectious agents likely to be

4 Wearing a face-mask, dissect the blood clots away using the operating microscope and watchmaker's forceps.③

5 Removing the connective tissue, transfer the portions of tissue required for culture to fresh HBSS.

6 Using fine forceps or sharpened needles, cut the tissue fragments, where applicable, into 300–400 μm cubes for adherent substratum-based cultures and up to 800 μm cubes for nonadherent explants.

7 To transfer the tissue fragments without contact, use minimum-displacement (Spemann) pipettes, transferring the explant in 10 μl HBSS (organ cultures).

8 To transfer the tissue fragments with precise positioning in a minimum fluid volume, use curved watchmaker's forceps. Pick up each fragment by the edge and carefully blot away HBSS from between the forceps tips with paper tissue, leaving 1–2 μl. Position the explant. Explants may be placed within a preformed collagen gel by this method.

9 Orientate the explant rapidly before the fluid evaporates, using mounted needles and with the aid of the operating microscope.

associated with the samples. When possible, attend the operation in person or arrange with theatre staff for the aseptic collection of the (carefully selected and labeled) samples and their prompt delivery to the culture laboratory. Principal sources of failure result from sepsis deriving from the tissue of origin (e.g. urogenital or gastrointestinal epithelia) or tissue anoxia caused by excessive delay prior to preparation.

② If dissection is to commence immediately (<15 min), the tissue may remain in an enclosed air environment. If a delay is expected, it should be transferred to MEM pre-gassed with 5% CO_2. Sample tubes should be kept on ice until dissection commences. All work should be carried out in a class II (minimum operational safety) cabinet dedicated to work with human tissue samples. The operator should wear gloves.

③ For preference, a microscope with a long-working distance (>25 cm for all magnifications) is desirable to reduce operator-derived sepsis. Ensure that all microscope surfaces are swabbed with 10% hibitane or 70% ethanol prior to commencing work.

Protocol 7. **Preparation of animal tissue explants.** L. Haynes

Reagents

Anesthetic (Fentanyl/fluanisone)⚠
Eagle's minimal essential medium (MEM) with penicillin and streptomycin (500 U/ml each), without serum
Hank's balanced salt solution (HBSS) without Phenol red
10% Hibitane or 70% ethanol⚠
Saline solution

Equipment

Class II microbiological safety cabinet
Electrolytically sharpened mounted needle
Glass evaporating basin
Macrodissection instruments
McIlwain tissue chopper or vibratome with sterile blades
Paper towelling
Petri dishes
Sterilized glass or disposable 'blunt' Pasteur pipettes cut back to the shaft (or long-handled medium forceps)
Sterile medium scissors
Sterile tubes
Syringe
Two 2 liter polypropylene beakers

Procedure

Method for adult animals and mammalian fetuses

1 Anesthetize the adult animal intraperitoneally or intramuscularly with fentanyl/fluanisone according to the manufacturer's directions. A terminal dose is normally given once the operation has been completed. A muscle relaxant may be given if skeletal muscle tissue is to be explanted. ①

Notes

This procedure takes approximately 2 hours for six explants, with dissection practice.

① Animal experimentation is controlled by legislation in some countries. Ensure that all necessary permissions are obtained in advance.

2 Sterilize the animal by total or partial immersion in 70% ethanol.

3 Dissect the tissue and remove to sterile tubes. If the tissue consists of a pregnant uterus, place segments with five fetuses each into separate sterile petri dishes.

4 Transfer the tissue vessels to the class II microbiological safety cabinet.

5 After dissecting a fetus from the uterus, transfer it to successive petri dishes of fresh saline to rinse free of cord blood.

6 Continue as in *Protocol 6*, or first mount and cut the tissues in a tissue chopper.

Method for avian embryos

1 In the preparation room, transfer eggs from the egg trays to a sterilized 2 liter beaker, swabbing each egg liberally with 70% ethanol.

2 In the class II microbiological safety cabinet, pierce the eggs with a mounted needle and rest one by one in the evaporating basin with the blunt end uppermost, supported by sterile paper towelling.

3 Cut a disc of shell off the top of each egg with sterile medium scissors and remove the embryo. Use 'blunt' Pasteur pipettes for embryos up to 6 days old *in ovo* or use long-handled medium forceps, with care, for eggs from 7 days *in ovo* onwards. Use the other hand, to cut the perivitelline membranes if necessary. Transfer the embryo to fresh HBSS.

4 Wash the embryo in successive petri dishes containing fresh HBSS to remove the yolk platelets.

5 Continue as for *Protocol 6*, or first mount and cut the tissues in a tissue chopper.

IV CORE TECHNIQUES. C. Morris, J.B. Griffiths, S. Warburton, C.M.L. West, M. Al-Rubeai, J.B. Clarke, F. Simione and A. Doyle

Routine subculturing: adherent cells. C. Morris

Adherent cell lines will continue to grow *in vitro* until either they have covered the surface available for growth or they have depleted the nutrients in the surrounding medium. At this point the cells must be subcultured into new vessels or they will rapidly die. A major feature of adherent cells is that they exhibit varying levels of adhesion depending on cell type; the majority of attached cell lines will require at least the addition of a protease to dissociate the cells from the growth surface.

It is important to observe some basic rules in order to optimize cell growth. Firstly the growth medium should be fresh and previously tested for maximum growth (see *Protocols 3–5*). The cultures must be handled in a safe and sterile environment (i.e. in a class II safety cabinet), which should ideally be placed in a room used only for cell culture. This type of cabinet provides protection to both the operator and the cells by ensuring that sterile air flows downwards across the working area, but does not escape into the room through the opening at the front. Antibiotics should not be routinely included in the medium for stock cultures. Antibiotics should only be used to remove contaminants in cultures that cannot easily be replaced. To prevent either microbial infection or cross-contamination of cell lines, it is essential to maintain separate reagent bottles for each cell line [i.e. phosphate-buffered saline (PBS), trypsin, medium]. Mark each bottle in use with the cell line name and date of use.

Additionally, each cell line must be handled separately, always working with those which have been characterized first. Allow a 10–15 min 'clearance time' between working with different cell lines, making sure that the flow cabinet has been thoroughly decontaminated by wiping all internal surfaces, including the sides and glass screen, and that all reagents for the previous cell line have been removed. If gloves are worn, these must also be changed, and the hands should be washed after each subculture.

Routine subculturing: nonadherent cells. C. Morris

Once established in culture, many tumor cell lines do not produce attachment factors and remain in suspension either as single cells or clumps of cells. Examples of this cell type are lymphoblasts and virus-transformed lymphocytes [4]. These are usually termed static suspension cultures because they do not need any mechanical movement to maintain the cells in culture. Alternatively, tumor or transformed cells which would normally grow as adherent cells, can often be maintained in suspension culture when continuously exposed to mechanical stirring (e.g. HeLa) [5,6]. Cells in suspension usually have very low levels of internal structural proteins (e.g. actin, vimentin) and lose some of their differentiated functions. However, suspension cell lines originally derived from attached cells will generally reattach if allowed to settle; returning to their original morphology and structural integrity.

Maintaining cells in active suspension (i.e. by means of a magnetically driven paddle or stirring rod), provides a more homogeneous environment than when they are grown as static cultures. Exchange of gases and nutrients becomes optimal due to the continual mixing of medium. Also, as environmental homogeneity can be maintained, the cells can be held in a steady state (homeostasis) for as long as fresh medium is available.

Cell quantification. J.B. Griffiths

Cell counting

To measure performance, to achieve reproducibility or to make comparative studies, a means of quantifying the cell population is needed. Classically, direct counts are made of cell numbers using a hemocytometer, usually in conjunction with a vital stain (e.g. Trypan blue) to distinguish viable and nonviable cells. However, all vital stains are subjective and cannot give absolute values. Nevertheless, the method is simple, quick and cheap, and requires only a small fraction of the total cells from a cell suspension. Automation of cell counting (total cells) is possible with electronic counters, especially for non-clumping single-suspension cells. A method which is widely used for determining total cell numbers is counting cell nuclei after dissolving the cytoplasm. This is particularly useful for large clumps of cells, where cells are inaccessible (e.g. in matrices). If determination of viability is the prime consideration, then the most accurate method is to count the cells which have the ability to divide by the mitotic index method.

Additional methods

It may not be possible to perform a direct cell count (e.g. during a monolayer culture or for cells immobilized in matrices) so an indirect measurement has to be carried out. Examples are glucose- or oxygen-consumption rates which can be converted to cell numbers from predetermined standard curves. Other metabolites include lactic and pyruvic acid and carbon dioxide. Again, there are reservations about such methods, as these metabolic rates are not constant throughout the growth cycle of a specific cell and may also be influenced by changes in the culture which are the subject of the investigation. However, they do provide a good indication of the viable cell mass in a culture, if not an absolute value, and are widely used to give growth yields or specific utilization rates. The advantage is, of course, that the method is nondestructive. A useful parameter increasingly being used is lactate dehydrogenase (LDH) activity release into the medium. This is released by dying/dead cells and therefore gives a quantitative measurement of the nonviable population. This method can be modified to measure viable cell mass by a reverse titration procedure; that is, by controlled lysis of the cells and measuring the increase in enzyme activity. Mention should be made of the use of radioisotopes to measure rates of synthesis (e.g. protein, DNA and RNA synthesis) [1]. However, these are usually kept for specific situations rather than general use.

There are many specific methods available which are suitable for a certain set of conditions or a particular experimental aim [8,9]. These include cellular ATP, intracellular volume, specific stains (e.g. MTT, Neutral red) redox potential, turbidity, calorimetry and biomass monitors.

General comments

The range of methods described have been selected to give quantitative measures of viable and nonviable cells (hemocytometer), viable and dividing cells only (colony-forming efficiency; CFE), viable (MTT and Neutral red) and loss of viability (LDH). The full range of methods are well documented in cell culture books (for examples, see refs 7–10).

Colony-forming efficiency. C.M.L. West

A clonogenic cell is defined as a cell with the capacity for sustained proliferation. Cells that can be shown to have undergone a minimum of five or six doublings to give rise to colonies containing at least 50 cells are said to be clonogenic. The ability to form colonies is a measure of reproductive integrity and is measured as colony-forming efficiency (CFE); it is often referred to as plating efficiency (PE) when colonies are grown on plastic. If colonies contain more than one cell type the appropriate terminology is colony-forming unit (cfu).

$$\text{cfu} = \frac{\text{number of colonies formed}}{\text{number of cells plated}} \times 100\%.$$

Clonogenic assays have a number of applications. Firstly, they can be used to reflect stem cell content (i.e. how many cells in a population are capable of unlimited proliferation). Secondly, they form the basis of assays for determining the lethal effects of cytotoxic agents. In the latter case a high degree of specificity can be obtained compared with growth or viability assays.

Statistics
For clonogenic assays where the total number of cells plated *in vitro* has been estimated Poisson statistics can be applied. In practice, if colony counts are small the Poisson distribution will be skewed, and when large (>10) the distribution will tend towards normal. Using the normal approximation to the Poisson distribution, the number of replicates required in an experiment to reduce the standard error to 5% of the mean can be calculated as, number of replicates $= 1/(0.0025 \times$ mean number of colonies per plate). In experiments with 100 colonies per plate only four replicates need be counted (i.e. a total of 400 colonies). If a standard error of 10% is acceptable, then the number of replicates required can be reduced by a factor of four. In practice this means that for experiments with 100 colonies per plate only one replicate plate is required, although in general three would be used to allow for inter-plate variability. In order to test whether two means are significantly different a

t-test can be used when the mean number of colonies counted is greater than 10 (i.e. the data approximate to a normal distribution). For mean values less than 10 (i.e. non-normal distributions) a χ^2 test or the nonparametric Mann–Whitney U-test should be used. An analysis of variance or the nonparametric Kruskal–Wallis test can be used for more than two groups. In order to establish the number of colonies that must be counted to obtain significant differences between two mean values, it has been estimated that a minimum of 2000 colonies should be counted for each mean value to make a 10% difference between values significant [11]. If the difference between the means is 50% then at least 100 colonies per group is required.

Multiplicity

$$m = \frac{\text{no. single cells} + 2 \times (\text{no. doublets}) + 3 \times (\text{no. triplets}) \text{ etc.}}{\text{no. single cells} + (\text{no. doublets}) + (\text{no. triplets}) \text{ etc.}}$$

Multiplicity correct counts should be used for clonogenic assays and it is important that multiplicity is kept to a minimum. If multiplicity is a problem, try using different disaggregation methods and times. The cells can also be filtered through sterilized nylon meshes. Colony formation does not distinguish between a single cell or group of cells since only one colony will arise.

The MTT assay. M. Al-Rubeai

The MTT assay [23] is a rapid, sensitive and quantitative assay that measures viability, proliferation and metabolic activation or inhibition of cells. The assay is based on the capacity of mitochondrial dehydrogenase enzymes to convert the yellow water-soluble substrate MTT into dark blue formazan product which is insoluble in water. The amount of formazan product is directly proportional to the viable cell number in a variety of cell types [23–28]. The results are consistent with those obtained from the [^{3}H]-thymidine uptake assay, although the MTT assay is more useful in that it detects cells which are not dividing but nonetheless are still metabolically active.

Cell cloning. J.B. Clarke

A clone is a population of cells which are all the descendants of one original cell. Clones may be derived both from continuous cell lines and from primary cultures. The object is to reduce as far as possible the genetic and phenotypic variation within a cell population.

Cloning cells from continuous lines has a number of applications. These lines are not necessarily genetically stable and their properties may alter during passage. Cloning may enable some cultures to be obtained which more closely resemble the original population. Variant cells can be isolated from continuous cell lines, especially after exposure to mutagens. Such cell types may include karyological variants, biochemical mutants and cells which exhibit different levels of product secretion or different susceptibilities to viruses. Transfection of cells with DNA does not necessarily yield populations with homogeneous genetic constitutions and biochemical properties, and cloning will enable cells to be selected and cultures developed with the required characteristics. In general, cloning is less successful with primary cells as these tend to have low CFE (colonies formed/single cells isolated).

Cell storage facilities and inventory. F. Simione

Storage facilities for low-temperature storage of cells can be set up using freezers that are cooled by mechanical refrigeration systems or with liquid nitrogen. Either system can be used for long-term storage of frozen cells as long as the criteria for assuring long-term stability are met. These criteria are: (i) maintaining a temperature that is below $-130°C$ at all times [29,30]; (ii) using an inventory system that allows access to desired material without compromising the temperature of stored material; and (iii) providing adequate safeguards for assuring that proper storage temperatures are maintained.

Cryopreservation. A. Doyle

At its most simple, crysopreservation [34] is based on slow freeze and fast thaw, together with high protein concentration and the presence of an agent which increases membrane permeability. The other, perhaps obvious, ingredient is a 'healthy' cell, although it has been known for back-up stationary phase cells to be cryopreserved in some laboratories with disastrous results.

The means of freezing at a controlled rate is a major consideration [35]. Dedicated controlled-rate freezing apparatus is considered by some as an expensive luxury. Nevertheless, successful freezing is virtually guaranteed (although it is always advisable to perform a viability check before discarding the live culture) and a chart recorder/printout of each freeze run can be kept. The 'traditional' method of a well-insulated box being transferred to −20°C for 24 h, then to −70°C for a further 24 h, then transfer to liquid N_2 can be used. Long-term storage at −70°C cannot be recommended.

Long-term storage

Long-term storage of cell stocks requires good record-keeping and maintenance of an inventory, with each removal logged to ensure adequate stock levels. This can best be achieved by the use of a microcomputer with appropriate software. A wide range of liquid nitrogen refrigerators is available; they must be alarmed to prevent accidental drying out.

Mechanical freezing

Cryopreservation using a −80°C or −130°C mechanical freezer is probably the least reliable of methods. However, low-temperature freezers are widely used to cryopreserve cell lines. In all cases the cells should be examined for viability following cryopreservation.

To achieve slow cooling, ampoules must be heavily insulated. A block of polystyrene with individual holes large enough to take ampoules should be prepared. The polystyrene must be 1–2 cm thick all around the ampoule, with no air spaces. The insulation required will depend on the final temperature of the freezer. The ampoules should be placed in the block, which is then positioned near the middle of the freezer and left overnight (16–24 h) before being transferred to a liquid nitrogen storage vessel. Before using this method on a regular basis, a series of tests should be carried out to check the cell viability after freezing.

Two-stage freezing

Ampoules are placed at a holding temperature of −20°C to −40°C for up to 24–48 h and then transferred directly to −196°C. It is essential to use freshly prepared freezing mixture every time. Dimethyl sulfoxide (DMSO) should be obtained from a supplier who is able to offer the most recently prepared stocks available. Once received this should be filter-sterilized (0.2 μm) through a filter specifically designed for DMSO (Nalgene) and stored in glass at −20°C. This will ensure that there are no problems due to oxidation of DMSO. In those cases where DMSO causes differentiation (e.g. the human promyelocytic leukemia cell line HL60) glycerol should be used.

Similar results can be achieved for a small number of ampoules by using a small device which holds ampoules in liquid nitrogen vapor in the neck of a Dewar flask. This device has a mechanism which allows the ampoules to be held at adjustable heights above the nitrogen, thus providing a variable cooling rate. After a holding time of 10–20 min the ampoules are plunged into liquid nitrogen and then transferred to their final storage location. Both of these alternatives require preliminary experimental work to determine the optimum conditions.

During the handling of frozen ampoules, a full-face protective mask and gloves must be worn. Ampoules containing pathogens should be thawed on an outer sealed container. In order to prevent explosion of ampoules on removal from the liquid phase it is recommended that transfer to the gaseous phase is carried out 24 h prior to thawing. To reduce the risk of injury from exploding ampoules during thawing it is advisable to place ampoules in a perforated container before placing in a water bath.

Adaptation of cells to suspension culture. C. Morris

Several historically important cell lines have been adapted to suspension culture from the originally adherent parent line. Examples are HeLa (human), BHK-21 (hamster) and L929 (mouse). The essential advantages to cells grown in suspension against adherent counterparts are:

- Proteolytic enzymes are not required for subculture.
- Simple subculture routines (i.e. dilution or centrifugation).
- Easy scale-up to large volumes.
- Homogeneity of culture nutrients.
- Single-cell suspension.
- High cell densities possible (i.e. up to 1×10^6–3×10^6/ml).
- High surface area to volume ratio (e.g. for introduction of agents to the cells).
- Total environmental control permissible once cells are capable of culture in bioreactors.

Methods available

Subculture of adherent cells using proteases (see *Protocol 8*)
The majority of adherent cell lines can be subcultured using the protease trypsin at a concentration of 0.25% (w/v). The chelating agent EDTA is often included to enhance the activity of the enzyme. Normally they detach individually to form a single cell suspension which can be counted and distributed into new culture vessels with fresh medium.
Problems: Slow cell detachment and aggregation may occur with some cell lines. This can be caused by a variety of factors: (i) trypsin in solution gradually loses activity at 4°C, and repeated warming and cooling also increases the loss of enzyme activity, (ii) cell aggregation or 'clumping' is due to surface membrane adhesion of carbohydrate

References

1. Health and Safety Executive (1988) *Control of Substances Hazardous to Health*. Approved Codes of Practice. HMSO, London.
2. Advisory Committee on Dangerous Pathogens (1990) *Categorization of Pathogens According to Hazard and Categories of Containment*. HMSO, London.
3. Kohler, G. and Milstein, C. (1975) *Nature* **256**, 495.
4. Steinitz, M., Klein, G., Koskimies, S. and Makela, O. (1977) *Nature* **269**, 420.
5. Gey, G.O., Coffman, W.D. and Kubicek, M.T. (1952) *Cancer Res.* **12**, 364.
6. Macpherson, I.. and Stocker, M. (1962) *Virology* **16**, 147.
7. Freshney, R.I. (1987) *Culture of Animal Cells – a Manual of Basic Technique*, 2nd Edn. Alan R. Liss, New York.

components produced during protease activity. To obtain a monodispersed culture, resuspend the cells in the smallest medium volume possible, and mix with a fine-tip pipette (e.g. glass Pasteur) or sterile yellow pipettor tip. Additional serum may also reduce the effect. If this is not successful, use EDTA (0.2 g/liter) dissolved in 0.85 g NaCl/liter or pronase to detach the cells. It should be noted that the enzyme activity of trypsin will be inactivated by the serum in the medium, but that for other proteases, such as pronase and dispase, this is not the case. All traces of these enzymes must be removed by pelleting the cells by centrifugation and then resuspending them in fresh medium prior to dispensing them into new flasks.

Mechanical removal of adherent cells (see *Protocol 9*)

Certain types of cell line may be harmed if exposed to proteases (e.g. NCTC 2071, L-M). Alternatively, some studies of surface membrane components cannot be performed in the presence of proteolytic enzymes. In such cases the collection or subculture of the cells requires mechanical removal from the culture vessel surface. The simplest method is to use a metal or glass rod bent at right angles and covered in a piece of silicon rubber tubing, sometimes called a 'rubber policeman'.

Problems: Mechanical removal of cells from a surface inevitably results in a loss of viability due to cell damage. It is also more difficult to produce monodispersed suspensions.

8. Patterson, M.K. (1979) in: *Cell Culture*, vol 11 (W.B. Jakoby and I.H. Paston, eds), pp. 141–151. Academic Press, New York.
9. Griffiths, J.B. (1985) in: *Animal Cell Biotechnology*, vol. 1 (R.E. Spier and J.B. Griffiths, eds), pp. 49–68. Academic Press, Orlando.
10. Doyle, A., Griffiths, J.B. and Newell, D. (eds) *Cell and Tissue Culture: Laboratory Procedures*. John Wiley & Sons, Chichester.
11. Hazout, S. and Valleron, A.J. (1977) *Cell Tiss. Kinetics* **10**, 569.
12. Courtenay, V.D. and Mills, J. (1978) *Br. J. Cancer* **37**, 261.
13. Courtenay, V.D. (1984) *Recent Results Cancer Res.* **94,** 17.
14. Hamburger, A.W. and Salmon, S.E. (1977) *Science* **197,** 461.
15. Bradley, T.R. and Metcalf, D. (1966) *Austr. J. Exp. Biol. Med. Sci.* **44,** 287.
16. Testa, N.G. (1985) in *Cell Clones: Manual of Mammalian Cell Techniques* (C.S. Potten and C. Hendry, eds), pp. 27–43. Churchill Livingstone, New York.
17. Puck, T.T. and Marcus, P.I. (1955) *Proc. Natl Acad. Sci. USA* **41**, 432.
18. Cox, R. and Masson, W.K. (1985) in *Cell Clones: Manual of Mammalian Cell Techniques* (C.S. Potten and C. Hendry, eds), pp. 170–174. Churchill Livingstone, New York.
19. Parkinson, E.K., Hume, W.J. and Potten, C.S. (1985) *Int. J. Rad. Biol.* **50,** 717.

Subculture of serum-free cultures (see *Protocol 10*)

Cells grown in serum-free culture medium are usually in a more 'fragile' state than their serum-grown equivalent and require gentle handling to prevent loss of viability.

Problems: Mechanical removal is not practicable, and enzymes present several problems as they tend to induce more damage to the cells due to endocytosis. If trypsin is used it will remain active because there is no serum present to neutralize it, and repeated washing in medium will not remove all traces. It is therefore necessary to add either some serum-containing medium to the cells, or a protease inhibitor and then wash the cells before subculture.

Subculture of nonadherent cells by dilution (see *Protocol 11*)

Cells that are grown either as static or stirred cultures will reach a maximum density dependent on the supply of nutrients and, in the case of mechanically stirred cells, the method of maintaining the cells in suspension. Normally this will be between 1×10^6 and 3×10^6 cells per milliliter. Once this density is reached the cells will cease to divide and they will start to die if the density is not reduced.

Problems: During growth, cells produce metabolic by-products (e.g. lactic acid) which become toxic if allowed to accumulate in the medium. Diluting cell cultures will initially remove sufficient waste metabolites to allow growth to continue, but after a few subcultures the viability of the cells will gradually decline. Surviving cells may

20. Rheinwald, J.G. and Green, H. (1975) *Cell* **6,** 331.
21. Rockwell, S. (1985) *Cancer Res.* **45,** 1601.
22. James, S.E., Arlett, C.F., Green, M.H.L. and Bridges, B.A. (1983) *Int. J. Radiat. Biol.* **44**, 417.
23. Mosmann, T. (1983) *J. Immunol. Methods* **65**, 55.
24. Al-Rubeai, M. and Spier, R. (1989) in *Advances in Animal Cell Biology and Technology for Bioprocesses* (R.E. Spier, J. Griffiths, J. Stephenne and P. Crooy, eds.), pp. 143–155. Butterworths, London.
25. Gerlier, D. and Thomasset, N. (1986) *J. Immunol. Methods* **94**, 57.
26. Grailer, A., Sollinger, H.W. and Burlingham, W.J. (1988) *J. Immunol. Methods* **107**, 111.
27. Niks, M. and Otto, M. (1 990) *J. Immunol. Methods* **130,** 149.
28. Sladowski, D., Steer, S.J., Clothier, R.H. and Balls, M. (1993) *J. Immunol. Methods* **157**, 203.
29. Meryman, H. (1966) in *Cryobiology* (H. Meryman, ed.), pp. 1–114. Academic Press, New York.
30. Mazur, P. (1966) in *Cryobiology* (H. Meryman, ed.), pp. 214–315. Academic Press, New York.
31. Mikoliczeak, J., Stephens, M. and Simione, F.P. Jr. (1987) *SIM News* **37**, 74.
32. Clarke, J.B. and Spier, R.E. (1980) *Arch.Virol.* **63,** 1.
33. Cooper, J.E.K. (1973) in *Tissue Culture – Methods and Applications* (P.F. Kruse and M.K. Patterson, eds), pp. 266–269. Academic Press, New York.
34. Ashwood-Smith, M.J. and Farrant, J. (eds) (1980) *Low*

also undergo selection pressures, resulting in altered characteristics. This method should only be used in the short term.

Subculture of nonadherent cells by sedimentation (see *Protocol 12*)
This involves separating the cells from the exhausted medium by centrifugation and inoculating them into fresh medium. By using this method, cell viability can be maintained indefinitely. New cultures are usually seeded at one-fifth to one-tenth of the maximum density reached (i.e. 1×10^5–3×10^5 cells per ml).
Problems: Inexperienced staff frequently over-dilute the cells, which will increase the lag phase considerably, and may even prevent the cells from dividing. Over-dilution can be remedied either by centrifuging the cultures and resuspending the cells in a smaller volume of medium, or by the addition of 10–15% of the exhausted medium to 'condition' the new medium.

Hemocytometer cell counts (*Protocol 13*)
The most common routine method for cell counting, which is both efficient and accurate, is to use a hemocytometer. This is a counting chamber 0.1 mm in depth, with squared graduations for ease of counting. Using a vital stain, such as Trypan blue, a viable and non-viable cell count can be made.
Problems: Accurate sampling, diluting and filling of the chamber are essential. Overfilling, inclusion of air bubbles and incomplete

Temperature Preservation in Medicine and Biology. Pitman Medical, London.
35. Doyle, A., Morris, C.B. and Armitage, W.J. (1988) in *Advances in Biotechnological Processes*, vol. 7 (A. Mizrahi, ed.), pp. 1–17. Alan R. Liss, New York.
36. Itagaki, A. and Kimura, G. (1974) *Exp. Cell Res.* **83,** 351.
37. Clark, J.B. and Spier, R.E. (1980) *Arch. Virol.* **63,** 1.

Protocols provided

8. *Subculture of adherent cells using proteases*
9. *Mechanical removal of adherent cells*
10. *Subculture of serum-free cultures*
11. *Subculture of nonadherent cells by dilution*
12. *Subculture of nonadherent cells by sedimentation*
13. *Hemocytometer cell counts*
14. *Viability studies*
15. *Determining the PE of an established adherent cell*
16. *Determining the CFE of a human tumor in soft agar*
17. *A soft agar clonogenic assay for anchorage-dependent cells*
18. *Determining the CFE of human lymphocytes using a limiting dilution assay*
19. *The MTT assay for cell viability*
20. *The spotting technique*
21. *Serial dilution*
22. *The cloning ring technique*

cleaning of the chamber causes errors. Inherent statistical errors are reduced by counting enough cells in duplicate determinations.

Cell viability studies (*Protocol 14*)
To ensure that a cell culture is growing exponentially it is useful to know the percentage viability and the percentage of dead cells. In Trypan blue, live healthy cells are usually round, refractile and relatively small in comparison with dead cells, which appear dark blue, larger and non-refractile when in suspension. Viability stains such as Trypan blue will only enter across the membranes of dead/nonviable cells.

Colony-forming efficiency (*Protocols 16–18*)
This is a true test of viability in that it depends upon a single cell being able to divide enough times to form a visible countable colony. A range of protocols is included to cover specific applications.
Problems: Dilution to isolate single cells is stressful to the cell, leading to cell death, thus additional feeding strategies, such as cell feeder layers may have to be used. Also, it takes many days to complete the assay.

Determining the CFE of a human tumor (*Protocol 16*)
Protocol 16 describes a method for plating four replicate tubes at a single cell density [12,13]. Where cell yields permit it is preferable to set up three different cell densities with eight replicate tubes.

23. *Preparation of feeder layers*
24. *Cryopreservation using controlled-rate freezing apparatus*
25. *Thawing frozen stocks of cells*
26. *Adaptation of adherent cell lines to suspension culture*

Another extensively used human tumor clonogenic assay employs a soft agar bilayer system on 35 mm petri dishes, alternative growth factors and growth in a 20% O_2 environment [14]. CFEs of human tumors can range from 0.001 to 5%.

Clonogenic assay for anchorage-dependent cells (*Protocol 17*)
Protocol 17 describes a method for determining the CFE of anchorage-independent cell lines and tumors, and hemopoietic cells. The CFEs of hemopoietic cells should be around 0.5–4% [15,16].

The MTT assay (*Protocol 19*)
This is a viability assay depending upon the activity of an enzyme which can be measured colorimetrically. It is quick, accurate and large numbers of samples can be tested, as it is an automatable method with a scanning spectrophotometer. It measures viable cells, whether dividing or nondividing.
Problems: Care is needed in the preparation of standard curves.

The spotting technique (see *Protocol 20*)
The spotting technique [32] depends upon the isolation of single cells in small spots of medium (volume approx. 1 μl) in wells of 96-well tissue culture trays. It is quick and convenient to perform, and allows a direct visual check that single cells have been isolated. It is very effective for cells with high CFE. However, if CFE is less than 5%, a large number of 96-well trays would be needed to generate a

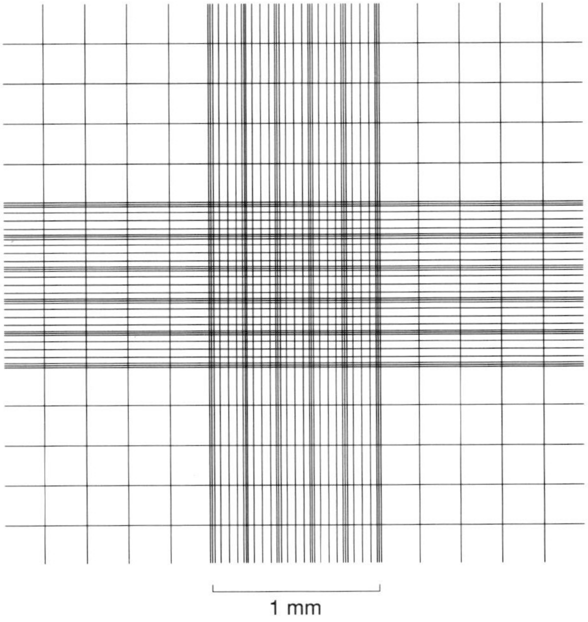

Figure 3. The hemocytometer showing a diagrammatic representation of areas to be counted.

significant number of clones, and the regular inspection of all the wells would be time-consuming and inconvenient. For this latter situation, the 'serial dilution' (*Protocol 21*) and 'cloning ring' (*Protocol 22*) techniques are more convenient, although it is easier with the 'spotting technique' to ensure that clones have each arisen from a single cell.

Problems: The small spots of cell-containing medium that are initially deposited may desiccate or undergo pH changes during microscopic inspection, before the addition of further medium. These risks may be reduced by setting up only part of a 96-well tray at a time to speed up the procedure, and by incorporating zwitterionic buffers (e.g. HEPES) into the culture medium. This technique is best suited to cell populations exhibiting CFEs of more than 10%. Below this, an inordinately large number of wells would have to be scanned to determine the presence of single cells.

Serial dilution (see *Protocol 21*)

In this procedure, volumes of serially diluted cell suspension are dispensed into the wells of 96-well tissue culture trays. As the cells grow, the trays are inspected microscopically to identify those wells which contain single colonies that appear to have arisen from a single focus of growth [33].

Problems: It may be difficult to be certain that a growing colony has arisen from a single focus of growth (i.e. a single cell).

The cloning ring technique (see *Protocol 22*)
In this procedure cloning rings of stainless steel, Teflon or glass, are used to isolate individual cell colonies growing in tissue culture-grade petri dishes or multi-well trays. The rings are sealed to the plastic surface with sterile silicone grease. Each colony may then be trypsinized individually and the cells removed for subculture.The procedure is also described in ref. 7.
Problems: Great care in manipulation is required at all stages. A complete seal between the base of the ring and the plastic surface is essential.

Protocol 8. **Subculture of adherent cells using proteases.** C. Morris

Reagents

Cell culture(s)
Culture medium
pH indicator
Sterile Ca^{2+}- and Mg^{2+}-free phosphate-buffered saline (PBS)
0.4% (w/v) Trypan blue in PBS⚠
0.25% Trypsin or 0.25% trypsin with 1 mM ethylenediamine tetraacetic acid (EDTA)

Equipment

Cell culture flasks
Class II microbiological safety cabinet
CO_2 incubator at 37°C
Hemocytometer
Microscope
Pipettes and pipettor

Procedure ①

1 Pre-warm the cell culture medium and PBS to the usual culture temperature. Allow the trypsin or trypsin/EDTA solution to reach room temperature. ②

2 Microscopically examine the cultures for their level of confluence, morphological appearance and any signs of microbial contamination.

3 Decant the culture medium from the cells by either pouring or pipetting it off. If the medium is poured off, in the case of flasks, invert the culture so that the medium runs down the side of the flask opposite the cells.

Notes

This procedure takes approximately 1 hour.

① ***Safety.*** Each cell line must be assessed according to its risk of infection (i.e. biohazard risk). In the United Kingdom hazardous substances, which includes biological material, are covered by COSHH regulations [1]. Animal cell lines which have not been fully characterized, should be handled under at least containment level II, unless a higher level is required (e.g. level III for hepatitis, human immunodeficiency virus- and simian immunodeficiency virus-infected cell lines [2]).

4 Wash the cells with a volume of Ca^{2+}-/Mg^{2+}-free PBS equal to at least half the volume of growth medium. Decant the PBS. Repeat, if the cells are known to be strongly adherent.

5 Add trypsin or trypsin/EDTA to the cells (1–2 ml per 25 cm^2 flask) and gently spread over the entire surface by tilting the vessel. Leave inside the safety cabinet for 15–60 sec, at room temperature, and then decant or pipette off most of the solution.

6 Place the cultures in an incubator at their normal growing temperature (e.g. 24°C for insect cell lines and 37°C for mammalian cell lines). The cells should detach after 2–10 min, as seen by gently tilting the vessel and observing whether the cell layer moves. Gently tap the side of the vessel to increase detachment. Detached cells will be seen as a 'milky' colored suspension.

7 Microscopically examine the cultures to ensure that all the cells have detached. Re-incubate if necessary.

8 Collect the cells by pipetting culture medium over the surface of the vessel (i.e. 2–5 ml/25 cm^2). Mix carefully to disperse the cells into a single-cell suspension. Measure the total suspension volume. ③

9 At this point the cells can be counted. When culturing a new cell line it is advisable to make a viable cell count to estimate the maximum cell density achieved (i.e. cells/cm^2 and the expected viability).

10 Transfer the cell suspension into prepared culture vessels by inoculating the appropriate volume of the cell suspension. For routine subculture a split-ratio is used (i.e. 1 into 2, 1 into 3, etc.). This is

② Repeated warming will inactivate trypsin.

③ Trypsin activity is neutralized by serum proteins. It is important to ensure that the volume of serum in the culture medium of the new culture vessels is equal to or exceeds the volume of trypsin in the vessel subcultured. Cells will not attach if the trypsin is not neutralized.

④ Alternatively, incubate the cells at a specific density (e.g. 2×10^4 cells/cm^2). First calculate the total number of cells required [i.e. total surface area (cm^2) × cell density required], and then the volume of cell suspension needed to inoculate the new vessels (i.e. total cells required divided by the number of cells/ml). For example, to inoculate a 75 cm^2 flask at 2×10^4 cells/cm^2 will require:

$$75 \times 2 \times 10^4 = 15 \times 10^4 \text{ cells } (= 1.5 \times 10^6)$$

If the suspension has 0.75×10^6 cells/ml, each new 75 cm^2 flask will require:

$$\frac{1.5 \times 10^6}{0.75 \times 10^6} = 2 \text{ ml of cell suspension.}$$

⑤ Cell lines vary in their plating efficiency (i.e. the number of cells in a viable population which will attach and grow). If the plating efficiency of a cell line is below 75%, care should be taken not to use high split-ratios which may reduce the optimum growth rate, due to a prolonged lag-phase.

calculated as the ratio of surface areas [e.g. a 75 cm^2 flask subcultured into 225 cm^2 (i.e. 3 × 75 cm^2 or 1 × 225 cm^2 Roux flasks) has a split-ratio of 1 into 3 (1:3)]. ④

11 Place the inoculated vessels in an incubator at the correct temperature. If a CO_2 incubator is used leave culture flasks caps loose. Check the cultures after a few hours for cell attachment and pH. If the cultures are too acid (yellow), or alkaline (purple), the CO_2 concentration must be checked immediately and adjusted. To prevent the loss of cultures due to a failure in the supply of CO_2, keep the flask caps tightened after the cultures have equilibrated (i.e. 1–2 h).

12 The duration between subcultures will depend on the incubation temperature and cell type. The majority of mammalian cell lines require subculturing every 3–7 days. If the duration is longer than 5 days, change the culture medium every 3–4 days. ⑤

13 For convenience, most attached cell lines are subcultured by dividing the contents of a vessel into an exact number of new vessels. However, if it is necessary to record the population doubling (pd) this can be calculated as follows:

two-fold split (1:2) = 1 pd
four-fold split (1:4) = 2 pd
eight-fold split (1:8) = 3 pd
16-fold split (1:16) = 4 pd.

This estimate is based on new cultures reaching the same cell density as the culture from which they were derived.

Protocol 9. **Mechanical removal of adherent cells.** C. Morris

Reagents

Cell culture(s)
Culture medium
Sterile Ca^{2+}- and Mg^{2+}-free phosphate-buffered saline (PBS)
0.4% (w/v) Trypan blue in PBS⚠

Equipment

Autoclave
Cell culture flasks
Class II microbiological safety cabinet
CO_2 incubator at 37°C
Hemocytometer
Micropipette and micropipettor
Microscope
Right-angled glass rod sheathed with silicon tubing

Procedure

1 Bag and sterilize the glass rods by autoclaving. A fresh rod should be used for each cell line.

2 Decant the culture medium and wash the cells with PBS.

3 Gently rub the flat edge of the angle across the cell sheet to remove the cells; inevitably some cells will be destroyed due to mechanical stress.

4 Collect the cells in a small volume of fresh medium and mix with a fine-tip pipette to break up any clumps.

5 Perform a viable count (*Protocol 14*) and calculate the number of cells required for subculture. Inoculate into new vessels.

Protocol 10. **Subculture of serum-free cultures.** C. Morris

Reagents

Cell culture(s)
Fetal bovine serum or soybean protease inhibitor
Serum-free culture medium
Sterile Ca^{2+}- and Mg^{2+}-free phosphate-buffered saline solution (PBS)
0.4% (w/v) Trypan blue in PBS⚠
0.25% Trypsin or 0.25% Trypsin with 1 mM ethylenediamine tetraacetic acid (EDTA) in PBS

Equipment

Cell culture flasks
Centrifuge and centrifuge tubes
Class II microbiological safety cabinet
CO_2 incubator at 37°C
Hemocytometer
Microscope

Procedure ①

1 After trypsinizing cultures (steps 1–7 in *Protocol 8*), collect the cells with fresh culture medium containing 10% serum (i.e. 3–5 ml/25 cm^2) and transfer into a sterile, screw-top centrifuge tube. ②③④

2 Centrifuge the cells at 80–100 *g* for 5 min.

3 Remove the supernatant and resuspend the cells in fresh serum-free medium.

Notes

This procedure takes approximately 30 min.

① Any trypsin present at the time of subculture must be removed or neutralized when cells are grown in serum-free medium.

② As an alternative to serum, sterile soybean trypsin inhibitor solution (Sigma, T 6522 and T 0800) can be added directly to the trypsinized cells at a concentration of 1 mg/1–3 mg trypsin used. After 5 min at room temperature the cells are centrifuged, the supernatant

4 Re-centrifuge the cells at 80–100 *g* for 5 min.

5 Remove the supernatant, resuspend the cells in a small volume of fresh serum-free medium and distribute into new vessels.⑤

removed, the cells washed twice in serum-free medium and then distributed into new vessels.

③ It may be necessary to precoat culture vessels to enhance cell attachment (e.g. collagen, dialyzed serum proteins).

④ When cell detachment is slow or incomplete or cell viability is low after detachment, it may be advisable to use an alternative protease [e.g. pronase and dispase (Boehringer-Mannheim Biochemicals) are bacterial proteases which are used to detach cell cultures]. Detachment is usually very rapid, but frequently with less damage to the cells. However, the enzymes must be removed by centrifugation as they cannot be neutralized with serum.

⑤ The plating efficiency of cells in serum-free culture will usually be lower than those in serum-containing medium, so the inoculating density may have to be increased to compensate.

Protocol 11. **Subculture of nonadherent cells by dilution.** C. Morris

Reagents

Cell culture(s)
Culture medium
Phenol red
0.4% (w/v) Trypan blue in phosphate-buffered saline (PBS)⚠

Equipment

Class II microbiological safety cabinet
CO_2 incubator or 5% CO_2, 95% air mixture
Hemocytometer
Low-speed magnetic stirrer
Mechanical stirrer vessels (e.g. Techne, Bellco, Wheaton) or plastic culture flasks
Microscope
Pipette tips and micropipettor

Procedure ①

1 Visually examine cultures for signs of high-density growth [i.e. cell turbidity and acid pH of the medium (orange to yellow when Phenol red is included in the medium)]. Static cultures should also be examined microscopically. Healthy cells will appear refractile, whereas dying cultures show cell lysis and shrunken cells. Check for signs of microbial contamination. ②

2 Remove a sample of cells by pipette for a viable cell count, ensuring that the culture is thoroughly mixed. If the cells have a tendency to clump, mix the sample again after removal, using a fine-tip pipette. To obtain a

Notes

This procedure takes approximately 30 min.

① For safety notes, see *Protocol 8*.
② If the cells usually grow in clumps, but are not in clumps at the time of examination this may be an indication of medium exhaustion.

statistically accurate cell count, at least 1% of the total culture volume must be removed for counting.

3 Perform a viable cell count by adding Trypan blue solution to the sample of cells in a ratio of between 1:2 and 1:10 (cells:Trypan blue) and counting with the hemocytometer (see *Protocol 13*).

4 Calculate the dilution required to obtain the inoculation density. For example, if the cell density is 1.2×10^6 and the new starting density is to be 2×10^5, the culture will have to be diluted six-fold:

$$\text{i.e. } \frac{12 \times 10^5}{2 \times 10^5} = 6$$

Calculate the new culture volume required, and add cells from the original culture to new culture medium (1 vol. of cells to 5 vols of medium in the above example). Where appropriate, gas with 5% CO_2. When a mechanically stirred vessel is used the speed should be set so as to just maintain the cells in suspension (i.e. 20–100 rev/min for Techne vessels and 30–150 rev/min for Bellco vessels; speed is increased with volume). Cultures grown in this type of vessel will reach maximum density faster than static cultures.

Protocol 12. **Subculture of nonadherent cells by sedimentation.** C. Morris

Reagents

Cell culture(s)
Culture medium
Phenol red
0.4% (w/v) Trypan blue in phosphate-buffered saline (PBS)⚠

Equipment

Centrifuge
Class II microbiological safety cabinet
CO_2 incubator or 5% CO_2/95% air mixture
Hemocytometer
Low-speed magnetic stirrer
Mechanical stirrer vessels (e.g. Techne, Bellco, Wheaton) or plastic culture flasks
Micropipette and micropipettor
Microscope
Sterile screw-top conical tubes (e.g. 15 ml, 50 ml)

Procedure①

1 Visually examine cultures for signs of high-density growth [i.e. cell turbidity and acid pH of the medium (orange to yellow when Phenol red is included in the medium)]. Static cultures should also be examined microscopically. Healthy cells will appear refractile, whereas dying cultures show cell lysis and shrunken cells. Check for signs of microbial contamination.

2 Remove a sample of cells by pipette for a viable cell count, ensuring that the culture is thoroughly mixed. If the cells have a tendency to clump, mix the sample again after removal, using a fine-tip pipette. To obtain a

Notes

This procedure takes approximately 30 min.

① For safety notes, see *Protocol 8*.
② To facilitate the calculation of cells required for inoculation adjust the numbers to the same factor (i.e. 10^5 or 10^6) so that the log values can be cancelled out.
③ Exceptions are slow-growing cells which must be maintained at lower densities (e.g. the mouse T cell, cytotoxic T lymphocytes) which must be kept in the density range 10^4–10^5 cells/ml.
④ Serum-free cultures. If low-serum or serum-free medium is

statistically accurate cell count at least 1% of the total culture volume must be removed for counting.

3 Perform a viable cell count by adding Trypan blue to the sample of cells between a ratio of 1:2 to 1:10 (cells:Trypan blue) and counting with the hemocytometer (see *Protocols 13* and *14*).

4 Calculate the volume of new culture(s) required, and prepare the culture vessels with medium. Where appropriate, gas with 5% CO_2.

5 Centrifuge the cells in sterile screw-top conical tubes at 80–100 *g* for 5 min. Calculate the number of cells needed to inoculate the new vessels using the equation: no. of cells × no. of ml of culture medium [e.g. If the inoculating density is to be 2×10^5 cells/ml and the new culture volume 500 ml, then the number of cells needed is $2 \times 10^5 \times 500 = 100 \times 10^6$ (1000×10^5).]

6 To calculate the volume of culture to be centrifuged, divide the total number of cells required by the number of cells/ml (culture density) (e.g. if the culture density is 1.2×10^6 cells/ml the volume of culture required for 100×10^6 cells is:

$$\frac{100 \times 10^6}{1.2 \times 10^6} = 83 \text{ ml}).$$ ②

7 After centrifugation, decant the supernatant, resuspend the cell pellet in a small volume of medium and inoculate into the new vessels. The majority of suspension cells can be maintained in the concentration range 10^5–10^6 cells/ml. However, when working with a new cell line keep

used with mechanically stirred cultures, it is important to control the culture pH. Bioreactors usually include this facility, but with simple vessels continued monitoring and addition of CO_2 is not always possible. Check cultures daily and re-gas when necessary. Reduction of the bicarbonate in the medium may facilitate cell growth, as may the addition of zwitteronic buffers (e.g. HEPES), as described in *Protocol 26*.

the cultures within a narrower range (e.g. 2×10^5–8×10^5 cells/ml) until their optimum density range can be determined. Once a growth curve has been established, maintain a density range which allows subculturing every 3–5 days. Shorter periods are inconvenient, and longer times between subculturing may result in nutrient exhaustion. ③

8 Re-gas culture vessels with 5% CO_2. This is especially important when using mechanically stirred vessels, as the pH of the medium has a tendency to rise in this type of vessel.

9 When a mechanically stirred vessel is used, the speed should be set so as to just maintain the cells in suspension (i.e 20–100 rev/min for Techne vessels and 30–150 rev/min for Bellco vessels); speed is increased with volume. Cultures grown in this way will reach maximum density faster than static cultures. ④

Protocol 13. **Hemocytometer cell counts.** S. Warburton

Reagents

70% Alcohol⚠
0.4% Trypan blue in phosphate-buffered saline (PBS) (passed through a 0.22 μm filter)⚠
0.25% w/v Trypsin/EDTA in PBS

Equipment

Hand-held counter
Hemocytometer with coverslip (Improved Neubauer Counting Chamber, BS 748: 1963)
37°C Incubator
Microscope
Pasteur pipette and pipettor

Procedure

1 Suspension cell lines can be counted directly. Gently wash attached monolayers twice with PBS. Add enough trypsin/EDTA to cover the monolayer and incubate at 37°C until the cells have detached (approx. 5 min). ①

2 Resuspend the cells in culture medium.

3 Thoroughly clean the hemocytometer and coverslip and wipe both with 70% alcohol before use.

4 Moisten the edges of the coverslip or breathe on the chamber to provide moisture before placing the coverslip centrally over the counting area and across the grooves.

Notes

This procedure takes approximately 30 min.

① The amount of PBS depends on the size of the cell culture flask (e.g. 5 ml for 1×25 cm^2).

② Newton's rings are rainbow-like interference patterns. The depth of the counting chamber is now 0.1 mm.

③ The central 1 mm square area is divided into 25 smaller squares each 1/25 mm^2. These are enclosed by triple-ruled lines and are subdivided into 16 squares, each 1/400 mm^2 (see *Figure 3*).

④ The following can cause inaccuracy of the hemocytometer method by affecting the volume of the chamber.
(i) Overflowing the chamber and allowing sample to run into the channels/moat.
(ii) Incompletely filling the chamber.

5 Gently move the coverslip back and forth over the chamber until Newton's rings appear, indicating that the coverslip is in the correct position to allow accurate counting. ②

6 Mix the cell suspensions gently but thoroughly and draw up a small sample into a Pasteur pipette to rest at the junction between the counting chamber and the coverslip. Draw in the cell suspension to fill the chamber. Both halves of the chamber should be filled to allow for counting in duplicate.

7 Use a light microscope at low power (×100 magnification) to focus on the counting chamber.

8 Count the cells in at least one of the 1 mm^2 areas. At least 200 cells should be counted to ensure an accurate count. This will determine how many squares should be counted. As a rule, the cells in the left-hand and top grid markings should be included in a square, those in the right and bottom excluded (see *Figure 4*).③

9 Count the cells in both halves of the chamber and calculate the number of cells/ml using the following equation:

$$\frac{\text{total number of cells counted}}{\text{no. of 1 mm}^2\text{ areas counted}} \times 10^4. \quad ④$$

(iii) Air bubbles or debris in the chamber.
There is an inherent error in the method due to random distribution of the cells in the chamber. The error is reduced by increasing the number of cells to be counted and duplicating the counts.

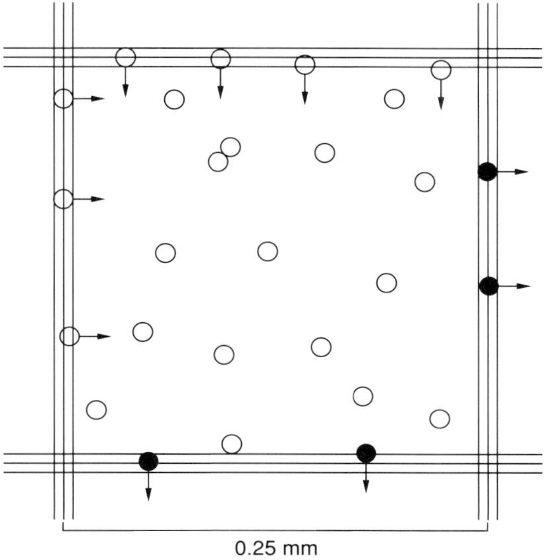

Figure 4. Diagram indicating which cells should be counted. ○ Count; ● count in the next grid.

Protocol 14. **Viability studies.** S. Warburton

Reagents

Cell culture
0.4% Trypan blue in PBS ⚠

Equipment

Hemocytometer
Microscope

Procedure

1 Dilute the cells 1:1 in Trypan blue. ①

2 Follow steps 1–8 from *Protocol 13*.

3 Count the viable cells in one or more of the mm^2 areas (N1, 2, 3 or 4).

4 Count the nonviable cells in the same areas. At least 200 cells should be counted to improve accuracy.

Calculations

(i) Total number of viable cells: $A \times B \times C \times 10^4$.

(ii) Total number of dead cells: $A \times B \times D \times 10^4$.

(iii) Total cell count: $A \times B \times C + D \times 10^4$ (= viable cell count + dead cell count).

Where: A = volume of cells, B = dilution factor in Trypan blue (step 1), C = the mean number of viable unstained cells, and D = the mean number of dead/stained cells. 10^4 is the multiplication factor used to convert 0.1 mm^3 to 1 ml.

Notes

This procedure takes approximately 40 min.

① Mix well and do not leave for longer than 1 min since Trypan blue is toxic to cells.

Protocol 15. **Determining the PE of an established adherent cell.** C. West

Reagents

Cell growth medium ①
5% Formalin (for human cells)
Gentian violet stain (1% in 70% ethanol) ②⚠
Typsin/1 mM ethylenediamine tetraacetic acid (EDTA)
2% Virkon

Equipment

5% CO_2/95% air incubator
Culture flasks
Inverted microscope
Petri dishes
Pipettes and pipettor

Procedure [17]

1 Trypsinize monolayer cultures or use cell suspension cultures and determine a viable cell count (see *Protocols 13* and *14*) and the multiplicity (see introductory text). If the multiplicity is greater than 1.05, vortex the cells or use a new flask of cells and a longer disaggregation procedure. It is essential that only single cells are plated.

2 Dilute the cells in growth medium to 100, 200 and 500 cells/ml (10 ml required).

3 Inoculate nine replicate petri dishes with 4 ml growth medium plus 1 ml cell suspension.

4 Place the plates in a humidified 5% CO_2 plus air incubator at normal growth temperature and rock the shelf or tray gently to and fro three

Notes

This procedure takes approximately 2 hours plus a 1–4 week incubation period.

① Eagle's basal medium (BME) plus 10% fetal bovine serum (FBS), 100 IU/ml penicillin, 0.1 mg/ml streptomycin and 2 mM glutamine. FBS should be neutralized with 2% Virkon for a minimum of 1 h before disposing of down a sink.
② In general, Gentian violet is quicker to use and stains colonies darker than either Methylene blue or Crystal violet. It is used as a 1% solution in 70% ethanol. Formalin (5%) can be added for human cells as a health and safety requirement. To use, aspirate the medium off the petri dishes; add 2–5 ml stain per petri dish; leave for 2–5 min; pour off the stain (this can be reused); rinse in tap water and leave the plates to dry.
③ If colonies are not evenly dispersed around the petri dishes,

times. The plates must not be moved now until the colonies have become stained. ③

5 Stain with Gentian violet and count three replicate plates per cell density at 1, 2 and 3 weeks (murine lines) or 2, 3 and 4 weeks (human lines). ④

6 Calculate the optimum cell densities for seeding and duration of incubation. ⑤

Keratinocytes and fibroblasts ⑥

Essentially the same method can be used by determining the CFE of human fibroblasts and keratinocytes using the appropriate culture medium [18]. The growth of fibroblast colonies requires the addition of heavily irradiated feeder cells (autologous or from another fibroblast line). Cells should be irradiated with 40–60 Gy X or γ radiation [19,20]. The growth of keratinocyte colonies requires the addition of heavily irradiated 3T3 fibroblasts (to inhibit fibroblast growth). For both fibroblast and keratinocyte colonies 14–16 days incubation are required. CFEs of established cell lines should be around 100%, and for human fibroblast and keratinocyte cultures around 10–40% and 1–5%, respectively. Fibroblast and keratinocyte cultures show a drop in PE with increasing culture life times. In general, it is better to use early-passage cells.

ensure that when the dishes are left on an incubator tray the tray is pushed up and down. Avoid circular movements as this will cause cells to settle in the middle of dishes.

④ Secondary colony formation can occur for rapidly growing cells. If colonies are surrounded by a large number of very small colonies ensure that petri dishes are not disturbed during incubation.

⑤ It is important to show linearity of colony formation (i.e. a plot of numbers plated vs. colonies formed should be linear over some range). If a threshold number of cells is required before colony growth occurs this indicates a requirement for feeder cells. The number of colonies formed will decrease if too many cells are plated. The number of colonies per culture that can be counted varies with cell type (e.g. 400 per 60 mm petri dish for tight human tumor lines vs. <50 for spread-out human fibroblasts).

⑥ Culture medium for human fibroblasts: DMEM supplemented with 10% FBS, 2 mM glutamine, 100 IU/ml penicillin and 0.1 mg/ml streptomycin. All lyophilized compounds (antibiotics, growth factors, hormones and stains) should be weighed out in a fume cupboard and gloves and masks should be worn.
Culture medium for human keratinocytes: 75% DMEM plus 25% Ham's F12 supplemented with 10% FBS, 10 ng/ml cholera toxin, 0.4 μg/ml hydrocortisone, 5 μg/ml transferrin, 5 μg/ml insulin, 10 ng/ml epidermal growth factor.

Protocol 16. **Determining the CFE of a human tumor in soft agar.** C. West

Reagents

August rat red blood cells ①
Culture medium ②
Iodonitrotetrazolium violet (INT stain) ③
Noble agar ④

Equipment

Class II microbiological safety cabinet
Falcon 2057 tubes
Low oxygen incubator (5% CO_2/5% O_2 in air)
Metal test tube rack
Microscope
Nylon filtration mesh (37 μm and 100 μm diameter pores)
Pipettes and pipettor
37°C and 4°C Water baths

Procedure

1 Disaggregate the tumor to obtain a single-cell suspension. Filter through a 100 μm filter and then a 37 μm filter. Obtain a total viable nucleated cell count (see *Protocol 13*) and determine the multiplicity. ⑤⑥

2 Place 0.5 ml cell suspension at 10× the required concentration (e.g. at 5 × 10^5 or 10^6 cells/ml) into a Falcon 2057 tube. Add 0.5 ml 1:8 diluted rat red blood cells.

3 Place 9.2 ml growth medium in a 37°C water bath for 10 min. Add 0.8 ml

Notes

This procedure takes approximately 4 hours plus a 4 week incubation period.

① Obtained by cardiac puncture. Wash three times remove the supernatant, resuspend in growth medium, incubate at 44°C for 1 h, dilute 1:8 in growth medium and store at 4°C for up to 1 week.
② Ham's F12 plus 15% fetal bovine serum, 2 mM glutamine, 2 μg/ml amphotericin, 25 μg/ml gentamicin, 10 μg/ml

5% agar (in practice 1 ml is added to allow for agar sticking to the sides of the pipette) and mix to give a final 0.4% agar solution.

4 Immediately add 4 ml of 0.4% agar to the cell-containing Falcon tube. The final agar concentration should be 0.33%. Pipette up and down twice to ensure an even mix. Aliquot 1 ml into the bottom of four replicate tubes.

5 Place the tubes in a rack at 4°C for a short period to ensure rapid agar setting. Once the agar has set, place the rack in a humidified 5% CO_2/5% O_2 incubator.

6 After 1 week, and weekly thereafter, add 1 ml growth medium to the cultures. On the third week remove 1 ml before adding 1 ml fresh medium. ⑦

7 At 4 weeks add 0.2 ml INT to each tube and incubate overnight. ⑧

8 Count the colonies more than 60 μm diameter using a microscope. ⑨

insulin, 2.5 μg/ml transferrin, 0. 5 μg/ml hydrocortisone, 10 ng/ml epidermal growth factor.

③ 0.5 mg/ml iodonitrotetrazolium violet in double-distilled water. Sterilize by autoclaving for 15 min. Store at 4°C. To use, warm the INT stain in a 37°C water bath for 10 min; add 0.2 ml stain to culture tube/petri dish and incubate overnight at 37°C; remove experiment (for human tumor work cultures can be fixed in formalin and placed in a fume cupboard overnight to release vapors). Cultures can be stored for at least 3 weeks at 4°C.

④ 5% solution in double-distilled water; sterilized by autoclaving or boiling 2 h. At the start of an experiment this must be melted on a hot plate and placed in a 50°C water bath until required.

⑤ Gloves must be worn when handling human material and all procedures should be carried out in a class II microbiological safety cabinet, including the loading of centrifuge holders. All human material and disposables in contact with human material should be disposed of by autoclaving, followed by burning.

⑥ See 'Multiplicity' in the introductory section. If the multiplicity is more than 1.05 and cannot be lowered, abandon this preparation. A single-cell suspension is crucial.

⑦ Low CFEs are generally a result of poor serum batch. It is very important to test serum to obtain a good batch. Sometimes the addition of heavily irradiated (40–60 Gy) feeder cells can be used to increase CFEs. Medium supplements (e.g. epidermal growth factor, insulin) can also be tried.

⑧ A positive control can be included to check for the presence of cell clumps (e.g. exposure to a high dose of a cytotoxic agent). A useful check for the presence of clumps and clusters in cell suspensions is to produce a radiation survival curve [21].

⑨ It is important to ensure that only colonies of more than 50 cells are counted. The appropriate minimum colony size varies with cell type and should be checked the first time that new cells are being studied (e.g. by counting the number of cells in colonies for colonies growing on plastic or by removing colonies from agar, flattening them with a coverslip on to a slide and then counting the number of cells).

Protocol 17. **A soft agar clonogenic assay for anchorage-dependent cells.**

C. West

Reagents

5% Noble agar ①

Equipment

Dissecting microscope

20% O_2 incubator②

35 mm Petri dishes

Test tubes

37°C Water baths

Procedure

1 Obtain a single-cell suspension and a total nucleated viable cell count (see *Protocols 13* and *14*).

2 Place 2 ml of cells at 5× the required plating density in a tube.

3 Warm 9.2 ml growth medium , including any required growth- or colony-stimulating factors, in a 37°C water bath. Do not warm for more than 10–15 min.

4 Add 0.8 ml 5% agar to the medium and immediately add 8 ml agar to the cell suspension. This will give a final agar concentration of 0.32%. The optimum agar concentration for each cell strain should be determined (usually in the range 0.2–0.4%).

Notes

This procedure takes approximately 3 hours plus a 3 week incubation period.

① 5% solution in double-distilled water; sterilized by autoclaving or boiling for 2 h. At the start of an experiment this must be melted on a hot plate and placed in a 50°C water bath until required. Alternative semi-solid support media are agarose and methylcellulose.

② Although the technique involves colony culture in 20% O_2, in many cases CFE can be improved by reducing this to 5%. If the incubator is opened frequently and the agar dries out, then the petri dishes can be placed in loose-lidded sandwich boxes with a small amount of sterile

5 Plate 1 ml cell–agar mixture into each of nine replicate petri dishes. Move each dish to and fro to ensure an even spread across the dish, with no air bubbles present. Leave the agar to set before moving the dishes.

6 Incubate the dishes at 37°C in 20% O_2. ②

7 Count the colonies on three replicate plates at 1, 2 and 3 weeks to determine the optimum incubation period.

8 Score the colonies under a low-power dissecting microscope. When the CFE of the cells is known, plate two or three cell densities (three or four replicates). If the CFE is unknown then three or four cell densities differing by factors of 10 should be plated.

distilled water at the bottom. Alternatively, six multi-well dishes can be used with four replicate cultures and two wells containing sterile water.

Protocol 18. **Determining the CFE of human lymphocytes using a limiting dilution assay.** C. West

Reagents

Culture media ①
HA15 phytohemagglutinin (PHA)
Human recombinant interleukin ②
Lymphoblastoid feeder cells (in Dutch RPMI + 20% fetal bovine serum) ②

Equipment

5% CO_2/20% O_2 incubator
Hemocytometer
Inverted microscope
Pipettes and pipettor
Sandwich boxes
60-well Terasaki dishes

Procedure [22]

1 Culture lymphocytes in growth medium overnight (5% CO_2) to remove adherent macrophages.

2 Prepare a feeder cell mix by combining:
- heavily irradiated (40 Gy) feeder cells (10^5 cells/ml);
- 20 IU/ml interleukin 2;
- 1% PHA. ③

Obtain a viable lymphocyte count using a hemocytometer (*Protocol 13*) and dilute the cells to 250, 500 and 1000 cells/ml.

Notes

This procedure takes approximately 2 hours plus a 2 week incubation period.

① Dutch RPMI (Imperial Laboratories) plus 9% FBS, 1% heat-inactivated human serum, 20 mM HEPES, 2 mM glutamine, 1 mM sodium pyruvate.
② Replenish stocks every 2 months.
③ These concentrations are double the final levels.

3 Add 1 ml lymphocyte cell suspension to 1 ml feeder cell mix and plate immediately into the 60-well Terasaki plates, with 20 μl per well.

4 Place the multi-well dishes into loose-lidded sandwich boxes and place the boxes in a humidified 5%CO_2/20% O_2 incubator at 37°C.

5 Read the plates at 10–14 days on an inverted microscope at ×40–×100 magnification. Wells are scored positive if 50 viable cells can be counted.

6 In order to obtain accurate measurements of CFE, 2–70% negative wells are required. The CFE is calculated by:

$$\frac{-\ln \text{no. of negative wells/total no. of wells}}{\text{no. of cells per well}} \times 100\%.$$

Protocol 19. **The MTT assay for cell viability.** M. Al-Rubeai

Reagents

1 ml 1M HCl (aq) added to 24 ml propan-2-ol (0.04 M HCl in propan-2-ol) ⚠
Cell culture
3-(4,5-Dimethylthiazol-2-yl)-2,5-diphenyltetrazolium bromide (MTT stain)⚠
Phosphate-buffered saline (PBS), pH 7.5
Phenol red

Equipment

0.22 μm Filter
37°C Incubator
Micro-ELISA reader
96-well Microtiter plate
Spectrophotometer

Procedure ①

1 Prepare a stock solution of 5 mg/ml MTT in PBS (pH 7.5) and filter through a 0.22 μm filter to sterilize and remove any small amount of insoluble residue.

2 To 100 μl of cell suspension or cell monolayer in each microtiter well add 10 μl of MTT (5 mg/ml) in PBS.

3 Cover the microtiter plate and incubate at 37°C for 2.5 h in a humidified incubator.

4 Add 100 μl of acid propanol to each well and mix thoroughly by pipetting 10–20 times to dissolve the blue formazan crystals. ②③

Notes

This procedure takes approximately 3 hours.

① Gloves and eye–face protection should be worn throughout the procedure as MTT is a carcinogenic and toxic agent. Reference should be made to hazard codes and instructions for protection.

② To reduce interference with the spectrophotometer reading by the Phenol red, remove culture medium before the addition of the acid propanol.

③ If acid propanol causes protein precipitation use instead 100 μl of 1:1 mixture of dimethyl sulfoxide and ethanol.

④ Plates should be read immediately after mixing, to avoid interference caused by precipitated proteins.

5 Read the plate on a micro-ELISA reader using a test wavelength of 570 nm and reference wavelength of 630 nm. ④

6 A linear relationship between cell number and MTT formazan exists up to 10^5 cells per well for a 2.5 h incubation with MTT (see *Figure 5*). For experiments involving the use of high cell density or for obtaining absolute cell number it is necessary that a standard curve be constructed by making a series of cell-doubling dilutions with 100 μl/well, using medium from the same culture. ⑤

⑤ The MTT assay can also be used to quantify cell activation by determining the maximal velocity (V) of the reaction. Alternatively, in order to compare two states of activation of a given cell, one can simply measure the MTT formazan produced by the same number of viable cells.

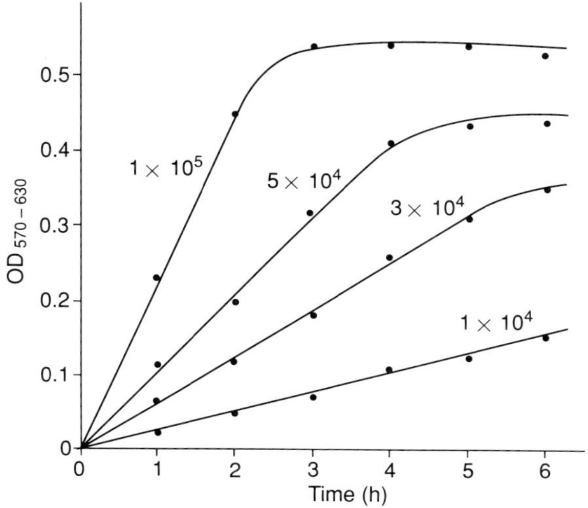

Figure 5. Relationship between hybridoma viable cell number and amount of formazan as affected by incubation time.

Protocol 20. **The spotting technique.** J.B. Clarke

Reagents

Cell culture
Complete medium including serum

Equipment

Class II microbiological safety cabinet
Inverted tissue culture microscope
Pasteur pipette
25 and 75 cm^2 Tissue culture flasks
96- and 24-well Tissue culture trays

Procedure

1 Resuspend adherent cells by the routine subculture method (*Protocol 8*). Use suspension cells directly from the culture medium. ①

2 Perform viable cell counts (*Protocol 14*) and adjust the cell concentration to between 500 and 1000 cells per ml.

3 Draw a small quantity of the cell suspension into the tip of a Pasteur pipette by capillary action.

4 Tap the tip of the loaded pipette against the base of each well of a 96-well tissue culture tray. This action deposits droplets of about 1 µl of suspension on the base of each well. The droplets should not touch the sides.

5 When all the wells of a tray contain a droplet, examine each well in turn

Notes

This procedure takes approximately 2 hours to set up, followed by several days incubation.

① Actively growing cultures give the highest CFEs.
② Occasionally, surface characteristics of the plastic bases of some wells may hinder full spreading of the droplets. This causes dark areas to appear around the periphery when viewed through the microscope. This also occurs when a droplet touches the side. Cells within such areas may often be seen by employing a higher magnification and/or adjusting the illumination. If, despite this, the whole of a droplet cannot be rigorously inspected, that well should not be used.
③ The incorporation of antibiotics into the medium during

for the presence of a single cell. Use ×50 magnification initially, then ×100 to confirm. Mark those wells that contain a single cell. ②

6 After inspecting each tray, add 200 μl of medium to each marked well containing a single cell. Alternatively, if deemed necessary, add 100 μl of complete medium and 100 μl of conditioned medium plus the resuspended feeder cells.

7 Incubate the trays at 37°C in a humidified atmosphere of 5% CO_2 in air. ③

8 Inspect the wells microscopically after 4 days and thereafter at 2 day intervals. When cell proliferation is observed, the wells may be fed at intervals with fresh medium. ④

9 When suitably sized colonies have formed (i.e. visible to naked eye), resuspend the cells by trypsinization or by gentle pipetting if nonadherent.

10 Subculture the contents of each well into a separate well of a 24-well microtiter tray. The 24-well tray may contain feeder cells and conditioned medium if considered necessary. The total medium volume per well should be 1 ml.

11 Incubate as before, and feed by replacing 50% of the growth medium every 4–7 days after growth commences.

12 When the cells cover 50–80% of the well subculture into 25 cm^2 tissue culture flasks. As growth proceeds, subculture into 75 cm^2 flasks.

13 Freeze down stocks of cloned cells over liquid nitrogen for long-term storage as early as possible.

the manipulative stage should be considered, as the cultures may have to be incubated over long periods of time, which will increase the contamination risk. However, there may be the possibility of some inhibitory or toxic effects and further long-term use is generally to be avoided.

④ Precise schedules will be governed by the observed behavior of the colonies, the known requirements of the cells, and characteristics of the media employed. Typically, 50% of the medium may be replaced every 4–7 days with vigorously growing cells.

Protocol 21. **Serial dilution.** J.B. Clarke

Reagents

Cell culture
Complete medium including serum

Equipment

37°C Humidified incubator with 5% CO_2/95% air
Inverted microscope
Pipettes and pipettor
25 and 75 cm^2 Tissue culture flasks
96- and 24-well Tissue culture trays

Procedure

1 Resuspend adherent cells by the standard subculture method (*Protocol 8*). Use suspension cells directly from culture.

2 Perform a viable cell count (*Protocol 14*).

3 For an expected CFE of less than 1%, prepare 10 ml of each of three cell suspensions with concentrations of 5000, 1000 and 500 cells/ml respectively; for a CFE of 1–10%, concentrations of 500, 100 and 50 cells/ml; and for over 10% CFE, 50, 10 and five cells/ml. ①

4 Add 100 μl of one cell suspension to each well of a 96-well tray. Repeat for each suspension, using separate trays.

5 The trays may contain feeder layers with conditioned medium if necessary. If these are not used, add 100 μl of complete medium to each well. ②

Notes

This procedure takes approximately 2 hours, followed by several days incubation.

① Actively growing cultures give the highest CFEs.
② Consider using feeder layers if the cells are known to have low CFEs or slow growth at low densities.
③ Feeding may disrupt nonadherent cell colonies, giving rise to new foci of proliferation. With these cells, therefore, feeding should be delayed until the operator is confident of the clonality of the colonies.

6 Incubate the trays at 37°C in a humidified atmosphere of 5% CO_2 in air.

7 Inspect the wells microscopically after 4 days and thereafter at 2 day intervals. Mark those wells in which a single colony appears. Ensure that there is only one center of growth as the colony develops.

8 When proliferation occurs, it may be advisable to feed the wells at intervals with fresh medium. The precise schedule will depend on the observed behavior of the colonies, the known requirements of the cells and characteristics of the medium in use. Typically, 50% of the medium may be changed every 4 days with vigorously growing cells.③

9 When suitably sized colonies have formed, resuspend the cells by trypsinization if adherent (*Protocol 8*) or careful pipetting if nonadherent. Subculture each clone into an individual well in a 24-well tray. The 24-well tray may contain feeder cells and conditioned media if considered necessary. Incubate as before and feed at 4–7 day intervals as the cells proliferate.

10 When sufficient cells are present (>50% of the area of the well is covered), subculture into 25 cm^2 flasks, and subsequently into 75 cm^2 flasks.

11 Master stocks should be stored over liquid nitrogen as early as possible.

Protocol 22. **The cloning ring technique.** J.B. Clarke

Reagents

Cell culture
Complete medium including serum
Silicone grease ①
0.25% Trypsin solution in phosphate-buffered saline (PBS)

Equipment

Cloning rings (glass, stainless steel or Teflon, approx. 0.5–0.8 cm inside diameter and 0.75 cm high) ①②
Four- or six-well tissue culture trays or tissue culture grade petri dishes
37°C Humidified incubator with 5% CO_2/95% air
Inverted microscope
Marker pen
Sterile forceps
Tissue culture flasks (25 or 75 cm^2)
24-well Tissue culture trays

Procedure

1 Seed cells into multi-well trays at 10 000, 1000, 100 and 10 cells/well. These quantities may be varied according to the expected colony-forming efficiency (CFE) of the cells and should be increased in proportion to the size of the container if petri dishes are used. The intention is to obtain discrete, well-separated colonies. Alternatively, set up dishes containing hybrid or recombinant cells under appropriate selective conditions. This will generally be done at the end of a hybridization or transfection procedure. The number of cells per dish will be governed by the proportion of hybrids or recombinants expected and their CFE in the selective media.

Notes

This procedure takes approximately 6–12 days.

① Sterilize the cloning rings and silicone grease by dry heat in separate glass petri dishes or foil-covered glass beakers.
② Stainless steel cloning rings are recommended. Their weight assists them to seat well and provide the necessary complete seal for trypsinization.

2 Incubate at 37°C in a humidified atmosphere of 5% CO_2 in air.

3 Inspect after 4 days and thereafter at 2 day intervals. Select colonies that appear to have arisen from one growth center and are well separated from other cells. Mark their location with a marker pen on the base of the dish.

4 Remove the medium from the cultures.

5 Using sterile forceps, pick up a cloning ring and dip its base in silicone grease. Ensure that the grease is evenly distributed.

6 Place the ring around the chosen colony, press down and move slightly to ensure a good seal. Repeat the procedure for the other selected colonies.

7 Fill the rings with trypsin solution. Leave for 20 sec, then remove, using a new pipette for each ring. Close the containers and incubate for 15–20 min at 37°C. Again using a new pipette for each ring, fill each ring with complete medium, carefully pump this up and down to dislodge the cells, and transfer the contents to an individual well in a 24-well tray.

8 Incubate at 37°C in an atmosphere of 5% CO_2 in air.

9 Inspect the wells at 4 day intervals. When sufficient cells are present (covering $>$50% of the base of the well), subculture into 25 cm^2 flasks and then into 75 cm^2 flasks.

10 Store master stocks of cells over liquid nitrogen at as early a stage as possible.

Protocol 23. **Preparation of feeder layers.** J.B. Clarke

Reagents

Complete medium suitable for chosen cells
Mitomycin C⚠
Phosphate-buffered saline

Equipment

Hemocytometer
37°C Incubator
Inverted microscope
25 cm^2 and 75 cm^2 Tissue culture flasks
Tissue-culture grade petri dishes.
96- and 24-well Tissue culture trays

Procedure

1 Grow the cells to about 50% confluency.

2 Add mitomycin C at a rate of 2 $\mu g/10^6$ cells or 2 $\mu g/ml$, either by adding stock solution in appropriate measured volume or by changing to fresh medium containing appropriate mitomycin C concentration.

3 Incubate for 24 h or overnight then wash the cells with four changes of prewarmed medium or phosphate-buffered saline. Add fresh medium and incubate for a further 24 h.①

4 Trypsinize the cells (*Protocol 8*) and resuspend in fresh medium at a concentration of 5 x 10^5 cells/ml.

Notes

① All mitomycin C must be removed (step 3) before the feeder cells are used.

5 Inoculate 0.1 ml per well into 96-well trays, 0.5 ml per well into 24-well trays or 5 ml per 25 cm^2 flask. Incubate for 48 h before use.

Modification for use of feeder layers in the 'spotting' technique

In the spotting' technique, the operator must visualize single cloned cells within the wells of 96-well trays. Clearly, the presence of pre-formed layers of feeder cells will prevent this. To circumvent this problem, proceed as follows:

1 After identifying wells containing single cells, add 0.1 ml of trypsinized feeder cells in fresh medium to each well, as in step 5 above.

2 Alternatively, resuspend the trypsinized feeder cells in fresh medium, inoculate 5 ml into each 25 cm^2 flask, and incubate at 37°C for 48 h. After this time, decant and retain the supernatant (conditioned) medium. Add 5 ml of trypsin solution to each flask, leave for 20 sec and discard. Incubate the flasks for 20 min or until cells are visibly detached. Add back 5 ml of the retained supernatant medium to each flask and agitate to resuspend the cells. Inoculate 0.1 ml of this suspension into each well.

Protocol 24. **Cryopreservation using controlled-rate freezing apparatus.**

A. Doyle

Reagents

Freeze medium [i.e. growth medium with 20% fetal bovine serum (FBS) and 9% dimethylsulfoxide (DMSO)] ①⚠

Equipment

Centrifuge
10 ml Centrifuge tubes
4°C Cooling chamber
Dispensing vessel
Drilled aluminum racks to hold ampoules
Inverted microscope
Liquid nitrogen refrigerator
1 ml and 10 ml Pipettes or a dispensing syringe (for larger volumes)
1.5–2.0 ml Plastic ampoules (Nunc)
Programmable freezer (Planer Products), two-stage freezer or well insulated polystyrene box
Protective goggles, face mask and gloves

Procedure

1 Resuspend adherent cells by the routine subculture method (see *Protocols 8, 11* and *12*). Suspension cells can be taken directly from culture. It is essential to ensure that any trypsin used for removal of adherent cells is neutralized by the addition of medium containing serum to a volume equal to that of the enzyme. ②

2 Perform a viability count by Trypan blue exclusion (*Protocol 14*). Viability

Notes

This procedure takes approximately 2–3 hours.

① In some cases 5–10% glycerol may be used. This is particularly important as DMSO causes differentiation in some cell types (e.g. HL60).

② Cell cultures must be healthy (confirmed by observation with an inverted microscope) and in log phase of growth. This can be ensured by using pre-confluent cultures at should be higher than 95%: a value below 80% is not usually acceptable and such cultures should be rejected.

3 Centrifuge the cells with the minimum of *g* force necessary to form a pellet (e.g. 100 *g* for 5–10 min).

4 Decant the supernatant and resuspend the cells in the cryoprotectant solution at a final concentration of 4×10^6–10×10^6 cells/ml. Dispense the cells into the ampoules using a pipette. For larger volumes an automatic syringe may be used but in this case the cell suspension should be placed on a magnetic stirrer in order to prevent settling. ③

5 Hold plastic ampoules at 4°C for up to an hour before freezing if several batches are to be prepared. The equilibration time can aid cell survival.

6 Transfer the ampoules for cryopreservation to the freezing chamber of a programmable freezer. The freezer should be programmed according to the manufacturer's instructions.

The following program is offered as one suitable for a wide variety of cell types. However, it is essential to test several combinations of cooling rate to optimize the viability on resuscitation. The main consideration is to minimize the effect of the latent heat of fusion during phase transition (i.e. when liquid freezes; see *Table 6*).

1 Ampoules should be held at 4°C for 10 min in the cooling chamber, prior to starting the freeze run. This allows them to equilibrate with the reference ampoule. ④

below maximum cell density and performing a medium change 24 h prior to freezing.

③ If large volumes of cell suspension are to be frozen, a problem can arise with bicarbonate-buffered medium due to increased alkalinity. It may be necessary to gas the mixture with 5–10% CO_2 in air to prevent cell death at pH values above 7.7. The use of zwitterionic buffers (e.g. HEPES) is not recommended as this may cause toxicity problems.

④ It may be necessary to adjust the temperature range and rate of cooling of ramp 2 to reduce the amount of heat generated during phase transition. The freeze medium will normally undergo this transition between −4°C and −8°C. Once the ampoules have reached −130°C, transfer to a nitrogen storage vessel (−196°C).

2 When the ampoules reach −60°C, the cooling rate can be increased to −10°C/min until −130°C is achieved.

Table 6. Temperature range and cooling rate

Ramp	Starting temperature (°C)	Upper temperature (°C)	Rate of climb (°C/min)
Ramp 1	Ambient	−10	−2
Ramp 2	−10	−30	−12
Ramp 3	−30	−60	−2
Ramp 4	−60	−130	−10

Protocol 25. **Thawing frozen stocks of cells.** A. Doyle

Reagents

Ampoules of frozen cells
1–2% Chloramine T ⚠
70% Ethanol ⚠
Growth medium
1–2% Virkon

Equipment

Culture flasks
Microscope
Pipettes
Tube rack
Water bath at 37°C

Procedure

1 Achieve rapid thawing by transferring ampoules directly to a 37°C water bath. If an ampoule contains potentially hazardous material it is advisable to add 1–2% (w/v) chloramine-T or Virkon to the water bath.

2 Take care not to submerge the cap of the plastic ampoule in order to prevent contaminated water entering the ampoule. A simple method is to use a plastic rack designed to hold tubes of the same diameter (e.g. Nalgene plastics 500 series) and place in the correct depth of water. Alternatively, a piece of foam with holes in it for the ampoules will float on the surface.

3 Once thawed, transfer to the sterile work area. Soak glass ampoules in 70% ethanol and allow to dry. A pre-scored ampoule can be snapped

Notes

This procedure takes approximately 30 min.

① Start-up cell density is of critical importance. The cells must be encouraged to condition the medium as rapidly as possible and prevent a protracted lag phase. In the case of most suspension cultures a cell density of 3×10^5–6×10^5 cells/ml is recommended. For adherent cells a density of 2×10^4–5×10^4 cells/ml is normally appropriate.

open using a tissue soaked in 70% ethanol. Unscored types must be cut with a diamond and then they can be soaked with 70% ethanol before opening. Plastic ampoules can be opened whilst being gripped with a tissue soaked with 70% ethanol.

4 Using a pipette, transfer the ampoule contents into either a flask or a centrifuge tube containing pre-warmed growth medium. Slow addition is recommended. If it is essential to remove the cryoprotectant at this stage, centrifugation should be at the lowest speed necessary to form a cell pellet. For cultures started directly from an ampoule the first medium change should be at 24 h unless recommenced otherwise. ①

5 Examine cells daily using an inverted microscope and subculture as soon as confluency or maximum cell density is reached. This is frequently indicated by a medium color change to orange or orange-red (i.e. acid pH).

Protocol 26. **Adaptation of adherent cell lines to suspension culture.** C. Morris

Reagents

Culture medium with serum
1.0 M HEPES buffer
Proteolytic enzymes (e.g. 0.25% trypsin/1 mM EDTA)
0.4% Trypan blue in phosphate-buffered saline⚠

Equipment

Autoclave
Centrifuge with swing-out rotor
Fume hood
Glass culture vessels with magnetic stirring bar
Hemocytometer
Incubator or 37°C hot room
Microscope
Silicon-treated (e.g. Sigmacote, Sigma Chemical Co.) culture vessel and mechanical stirrer

Procedure

1 Glass culture vessels must be siliconized to prevent cell attachment (e.g. Sigmacote). Treatment requires the use of a fume hood due to toxicity. Manufacturer's instructions should be followed carefully when handling these solutions. After treatment the vessel should be rinsed with sterile water and sterilized by autoclaving. The vessel can normally be used 3–10 times before re-siliconizing is required. The simplest culture vessel for suspension cells is a glass bottle with a Teflon-coated magnetic bar. A better design is a magnetic bar suspended from a rod attached to the lid (e.g. Bellco vessel). Both vessels will, however, require a mechanical stirrer with a low speed capability (i.e. 20–150 rev/min). ①

Notes

This procedure takes approximately 10–20 days.

① A system developed by Techne provides both a special stirrer vessel and an electronic speed controller which provides reliable slow-speed control. This is a particularly important aspect of this type of specialist equipment. Temperature control should be to +/−0.5°C.

② The time required to adapt an adherent cell line to suspension can only be discussed in general terms. In the case of some cell lines it is possible to maintain a suspension culture in which little or no further growth occurs but the

2 Prepare cells using the appropriate proteolytic enzyme(s) and once a monodispersed suspension has been produced, centrifuge the cells in serum-containing medium with at least 1 ml of 10% serum-containing medium per 2×10^6 cells. Serum is essential to neutralize enzyme activity.

3 Resuspend the cell pellet in fresh growth medium and recentrifuge after taking a sample for viable cell counting.

4 Resuspend the cells to a density of 2×10^5 and 4×10^5 cells/ml in 10% serum-containing medium then transfer to a sterile siliconized vessel. Initial adaptation should involve small volumes (i.e. 25–200 ml) and the medium should be gassed with 5% CO_2 if required. It may be possible to use the same medium for suspension cells as for the attached cells. However, the control of pH is important, as there is a tendency towards a high pH because of the greater oxygen interchange between the gas and liquid phases during mechanical stirring. This can be overcome by reducing the bicarbonate content of the medium and including a zwitterionic buffer (e.g. 10–25 mM HEPES). To maintain the correct osmotic balance, the NaCl content of the medium should be reduced by 7% when 20–25 mM HEPES is included. Recommended levels of bicarbonate and HEPES are given in the table at the end of this protocol. ②③

5 Set the stirring speed at 25–100 rev/min, depending on cell volume and the type of vessel. The speed should be sufficient to just maintain the cells in suspension. Too high a speed will damage the cells.

6 Count the cells daily (i.e. total cell count and viability). When the cells reach a density of 8×10^5–9×10^5/ml they can be subcultured. ④⑤

cells remain viable for a period of 5–20 days. In general, the more strongly adherent and specialized in function a cell is, the less likely it is to adapt easily. Better candidates are highly transformed or tumorigenic lines. An indication of whether a cell is capable of growth in suspension is seen if they detach easily during mitosis and produce a number of floating cells in the culture.

③ Although the inclusion of HEPES should allow the omission of CO_2, the majority of cells will not grow well in the complete absence of CO_2 as shown in the study by Itagaki and Kimura [36]. However, the inclusion of a zwitterionic buffer will obviate the effects of temporary fluctuations in CO_2 tension. An initial serum concentration of 10% should be used but may be reduced once cells adapt. Reduced serum concentration will, however, enhance the pH gradient.

④ Initially adaptation may reduce the viable cell population in the culture. A continued decline will indicate that either the cells cannot adapt or that one or more environmental parameters is not optimal. Reduction of the stirring speed, increasing the serum concentration and monitoring the culture pH should be examined as first possibilities. Protective agents such as pleuronic F-68 (BASF) can be included to minimize the shearing forces during stirring.

⑤ Cultures with good viability but exhibiting slow cell division have retained surface attachment factors, which inhibit growth while the cells remain in suspension. A period of several weeks may then be necessary for these

7 Remove a volume of cells sufficient to inoculate a new vessel at 2×10^5–3×10^5 cells/ml and centrifuge them at 70 *g* for 10 min.

8 Resuspend the cells in a small volume of medium and inoculate the new vessel(s).

9 Once growth and a steady viability have been established, the cultures can be expanded and cells prepared for cryopreservation as a master (seed) cell bank. The master cell bank should be used to prepare working cell banks, in order to retain cell characteristics. ⑦

Gas phase CO_2 (%)	$NaHCO_3$ (mM)	HEPES (mM)
Atmospheric	4	10
5	25	20
10	40–45	30–45

cells to lose their dependency. If, however, cell growth does not improve within 10–20 days, it is unlikely that the cells can be adapted to suspension in culture.

⑥ During preparation with enzymes, cells may form large aggregates due to surface modifications of cell membrane glycoproteins. If these cell clusters remain after washing in serum-containing medium resuspend the cells in 0.2 g/liter EDTA in NALCO solution (0.85 g/liter) and incubate at 37°C for 5–15 min. After mixing with a fine-tip pipette, a homogeneous suspension should be produced. Pellet the cells by centrifugation at 70 *g* for 10 min and transfer to a culture vessel. If the problem persists in culture, it may be necessary to reduce the serum concentration and include a low level of EDTA in the first few cultures at an initial concentration of between 0.01 and 0.02 g/liter.

⑦ An alternative approach is to select clones for adaptation. This is especially useful when specific properties are required (e.g. virus susceptibility, protein secretion). A method is described by Clarke and Spier [37] for adaptation of BHK cells, using microtiter trays.

V CORE CULTURE SYSTEMS. B. Griffiths

Introduction

The basic culture unit for animal cells is a tissue culture flask made of special quality polystyrene in sizes ranging from 5 cm^2 to 1500 cm^2 by a wide range of manufacturers. These are ideal for anchorage-dependent cells and they will also support a wide range of suspension cells, although for some cells either a gentle orbital shaking device for the flasks, or a shake flask is needed. The use of these, and analogous, basic culture units is well documented [1,2].

The techniques in this section are aimed at the next level of operation (i.e. the first scale-up step). To put a perspective on scale-up, industrial scale processes are operating at over 10 000 liters which, from a 10 ml start-up culture, is a million-fold scale-up. The processes described in this section are traditionally below 10 liters, although all of them are used in manufacturing processes (either by increasing the number of units, or volume).

One aim in scale-up is to move from multiple-unit processes (based on flasks and roller bottles) to a high-volume (and/or high cell density) unit process. Thus the choice of methods encompasses this aim by describing roller bottles (a multiple process), and spinner flasks and microcarrier culture, which are the basic unit processes for suspension and anchorage-dependent cells, respectively. There are a large number of alternative culture systems available [3–5] as a result of: (i) the different needs of suspension and anchorage-dependent cells, and (ii) multiple solutions to overcome growth-limiting factors. These factors are oxygen supply (cells are damaged by high oxygen transfer conditions of sparging and fast stirring rates), increasing surface area in a space-efficient three-dimensional system (of which the miocrocarrier is the most effective method), nutrient limitation (particularly glutamine and glucose) and toxic metabolite build-up (especially ammonia and lactic acid or pH), and the scale-up logistics of handling large quantities of cells and media, in a timely and aseptic procedure [6].

Suspension culture is the preferred method over monolayer for scale-up because it is easier and cheaper, requires less space, facilitates sampling and thus measurement of cell growth and the control of environmental factors. However, anchorage-dependent systems do have the advantage of allowing easy nutrient supplementation by media changes, being good for perfusion (should high-density growth be needed), and facilitating changing the cell/medium ratio (useful for cell attachment, viral infection, enzyme treatment, etc.). Microcarrier culture has the advantage that it has most of the benefits of both suspension and anchorage-dependent systems.

Some key factors for the successful scale-up of animal cells include: inoculating cells in prime condition (harvested from late logarithmic growth, not stationary phase); to harvest and re-inoculate as quickly as possible, especially minimizing exposure time to proteolytic enzymes (e.g. trypsin); avoiding over-robust mixing of post-trypsinized cells; inoculating into fully equilibrated medium, as shifts in pH, temperature and oxygen levels during culture initiation are very damaging; and ensuring that nutritional, physiological and physicochemical conditions are optimized.

Methods available

Roller bottle culture (see *Protocol 27*)
The first scale-up step for anchorage-dependent cells is from a stationary flask to a cylindrical bottle which is slowly revolved to bathe the cell sheet growing on the entire internal surface with nutrient medium [8].
Problems: Large bottles (>1500 cm^2) are unwieldy to manipulate in laminar flow cabinets. Some cell lines are well suited to roller bottles but others, particularly epithelial cells, give problems of streaking, clumping and inadequate spread over the total surface. This is due to cells being nonlocomotory (the rate of attachment is also a factor).

References

1. Freshney, R.I. (1987) *Culture of Animal Cells. A Manual of Basic Technique.* Wiley-Liss, New York.
2. Griffiths, J.B. (1992) in *Animal Cell Culture: a Practical Approach* (R.I. Freshney, ed.), pp. 47–93. IRL Press, Oxford.
3. Griffiths, J.B. (1988) in *Animal Cell Biotechnology*, Vol. 3 (R.E. Spier and J.B. Griffiths, eds), pp. 179–220. Academic Press, London.
4. Prokop, A. and Rosenburg, M.Z. (1989) *Adv. Biochem. Eng./Biotechnol.* **39**, 29.
5. Glacken, M.W., Fleischaker, R.J. and Sionsky, A.J. (1983) *Trends Biotechnol.* **1**, 102.
6. Griffiths, J.B. (1990) *Cytotechnology* **3**, 109.

Spinner flask culture (see *Protocol 28*)
The first scale-up step for suspension cells is the spinner flask, developed from the concept of putting a magnetic bar in a bottle and placing this on a magnetic stirrer by the need to have the magnet suspended (to avoid damaging cells) and a geometry which permits efficient mixing [9]. It is a glass vessel in the size range of a few milliliters to 10 liters placed on a multi-based magnetic stirrer in a hot room.
Problems: If cells are prone to clumping or adhering to glass surfaces, then medium with a reduced Ca^{2+} and Mg^{2+} concentration should be used. Siliconization of the glass vessels (e.g. with Repelcote or Sigmacote) is good practice.

Microcarrier culture (see *Protocol 29*)
Microcarriers are small particles, usually 100–300 μm in diameter, on which cells will attach and grow and which can be suspended in stirred culture medium [10]. The value of the method is that culture vessels already in use for suspension cells can be used for microcarriers with little process modification, and that it provides an enormous unit surface area (12 cm^2/ml: i.e. 1 liter gives equivalent cell capacity to 8–15 roller bottles).
Problems: Adherence to proven working procedures is critical as this is a method which has little latitude in experimental conditions so care, planning, and the right tools must be used.

7. Griffiths, J.B. (1995) in *Cell and Tissue Culture; Laboratory Procedures* (A. Doyle, J.B. Griffiths and D.G. Newell, eds), 28:0.1–28:0.7. John Wiley & Sons, Chichester.
8. Griffiths, J.B. (1995) in *Cell and Tissue Culture; Laboratory Procedures* (A. Doyle, J.B. Griffiths and D.G. Newell, eds), 28A1.1–28A:1.4. John Wiley & Sons, Chichester.
9. Griffiths, J.B. (1995) in *Cell and Tissue Culture; Laboratory Procedures* (A. Doyle, J.B. Griffiths and D.G. Newell, eds), 28A:2.1–28A:2.5. John Wiley & Sons, Chichester.
10. Griffiths, J.B. (1995) in *Cell and Tissue Culture: Laboratory Procedures* (A. Doyle, J.B. Griffiths and D.G. Newell, eds), 28C:1.1–28C:1.7. John Wiley & Sons, Chichester.
11. Looby, D. and Griffiths, J.B. (1990) *Trends Biotechnol.* **8**, 204.
12. Griffiths, J.B. (1990) in *Animal Cell Biotechnology*, Vol. 4 (R.W. Spier and J.B. Griffiths, eds), pp. 149–166. Academic Press, London.
14. De Bruyne, N.A. (1988) in *Animal Cell Biotechnology*, Vol. 3 (R.E. Spier and J.B. Griffiths, eds), pp. 142–178. Academic Press, London.
15. van Wezel, A.L. (1967) *Nature* **216**, 64.
16. Griffiths, J.B. (1991) *Chem. Indust.* **18**, 682.
17. Griffiths, J.B. (1990) in *Methods in Molecular Biology*, Vol. 5 (J.W. Pollard and J.M. Walker, eds), pp. 47–93. Humana Press, Clifton, NJ.

Choice of method

Microcarrier culture can be used for both anchorage-dependent and suspension cells (if porous carriers are used) [11,12], is scaleable as a unit process to many thousands of liters, is easy to monitor and control environmentally, easy to sample, and is good for many culture manipulations (e.g. medium changes, perfusion, concentrating or diluting cell number, etc.).

Roller bottle culture

The advantage of a roller bottle over stationary flasks is that the whole of the internal surface area is utilized, a smaller volume of medium is used per unit surface area and thus a higher titer of product is obtained; cells are more efficiently oxygenated due to alternating exposure to medium and gas phase and, being a dynamic system, better mixing reduces concentration gradients and supports higher unit cell densities [8]. The basis of the system is to place multiple cylindrical bottles (glass or plastic) into an apparatus which will rotate the bottles evenly at set rotational speeds (between 5 and 60 rev/h). Apparatus is available to accommodate from four to several hundred bottles in the range of 250–2000 cm^2 capacity.

A multiple process requires considerable staff time for repeated manipulations. Thus, many modifications have been introduced that

18. *Microcarrier Cell Culture – Principles and Methods*. Pharmacia LKB Biotechnology 5-75182, Uppsala, Sweden.
19. Griffiths, J.B., Cameron, D.R. and Looby, D. (1987) *Devel. Biol. Standard*. **66**, 331.
20. Sanford, K.K., Earle, W.R. and Evans, V.J. (1951) *J. Natl Cancer Inst*. **11**, 773.
21. Crespi, C.l. and Thilly, W.G. (1981) *Biotechnol. Bioeng.* **23**, 983.
22. Reuveny, S. (1990) in *Large-scale Mammalian Cell Culture Technology* (A.S. Lubiniecki, ed.), pp. 271–341. Marcel Dekker, New York.

Protocols provided

27. *Roller bottle culture*
28. *Spinner flask culture*
29. *Microcarrier culture*

increase the efficiency or surface area capacity of roller bottles. Some examples are as follows:

(i) Perfusion. Specially modified swivel caps have been developed that allow continuous perfusion of roller bottles (e.g. Bellco Autoharvester, New Brunswick apparatus) [13].
(ii) Spira-Cell multi-surface roller bottle (Sterilin Ltd). This bottle contains a spiral polystyrene cartridge on which cells grow on both sides of the film; 3000 cm^2, 4500 cm^2 and 6000 cm^2 sizes are available.
(iii) Extended surface area roller bottle (ESRB) (Sterilin Ltd). This polystyrene bottle has a ribbed or corrugated surface, doubling the surface area of the 850 cm^2 flask to 1700 cm^2.
(iv) Textured silicone rubber-coated bottle (Textur Sil, Ashby Scientific Ltd).
(v) Bellco-Corbeil Culture System (Bellco). A roller bottle is packed with a cluster of small glass tubes (arranged in parallel and separated by silicone spacer rings). Models are available in 1000 cm^2, 10 000 cm^2 and 15 000 cm^2 sizes. Medium is perfused through the vessel and the bottles rotate 360° in alternate directions to avoid tube twisting and the need for special caps.
(vi) Mini PERM (Heraeus Instruments) suitable for suspension cells such as hybridomas.
(vii) The rotary cell culture system (RCCS) enables the cultivation

of differentiated three-dimensional culture to mimic parent tissues due to a microgravity effect. This is therefore not a scale-up system and is available in volumes from 10 to 250 ml.

An alternative scale-up route to roller bottles is to use multi-surface stationary systems such as the Cell Cube (Costar) or Cell Factory (Nunc).

Spinner flask culture

A typical spinner flask is shown in *Figure 6a* and has a stirrer shaft containing the magnet; side arms for adding cells, changing the medium, gassing with O_2- or CO_2-enriched air (0.22 μm filters must be attached); and a slightly convex base under the stirrer to aid stirring. There are variations on the theme from different suppliers (e.g. *Figure 6b*, an asymmetric stirrer with a vertical magnet [14]). High-quality stirrers which give stable speeds in the range 10–300 rev/min, do not need re-setting after interruptions in power, have accurate tachometers and do not generate heat on the top surface are essential.

There is a wide range of spinner vessels available:

(i) Conventional vessels (*Figure 6a*) from Bellco, Wheaton, Celltech.
(ii) Conventional vessels but with large-bladed paddles attached to the magnet or base of the stirrer shaft (Bellco and Cellon

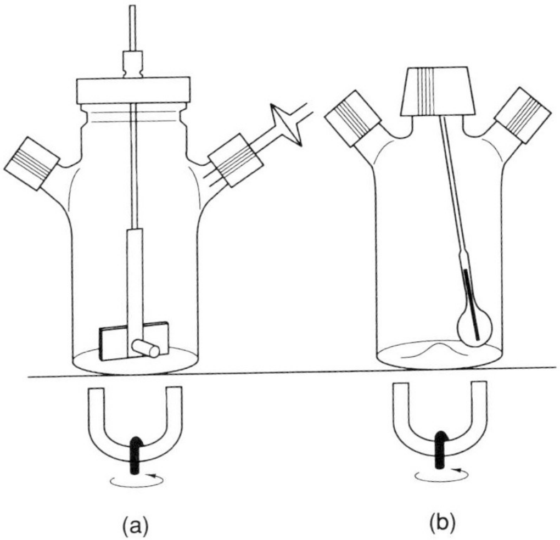

Figure 6. (a) A typical spinner flask with a magnetic stirrer bar. (b) A Techne flask with an asymmetrical stirrer rod.

microspinner flasks). These vessels give more efficient stirring with better mixing at the lower speeds and are thus primarily aimed at microcarrier culture.

(iii) The Techne system uses a radial stirring action which is designed for good mixing at low stirring speeds and to minimize mechanical stress to cells (*Figure 6b*) [14].

(iv) The Technomara cell spin flask with a double pendular oscillation system.

(v) Braun Super Spinner with perpendicular stirring system of porous membrane material.

(vi) The Techne BR-06 Bioreactor uses a floating impeller which again gives a gentle stirring action. An advantage of this system is that the vessel can be used equally effectively at various working volumes from 500 ml to 3 liters. This allows an *in situ* build-up of cell seed by just adding extra medium as cell growth occurs.

If both side arms are fitted with filters, then continuous aeration through the head space can be carried out, giving better control of pH and higher cell densities.

Microcarrier culture

The growth of anchorage-dependent cells on small particles (100–300 μm) suspended in a stirred culture system has made a tremendous impact on upstream processes since its conception in 1967 by van Wezel [15] (for reviews see refs 6,16,17). A wide range of microcarriers have been developed (*Table 7*) based on dextran, gelatin, collagen, polystyrene, glass, cellulose, polyacrylamide and silica which will accommodate virtually all cell types. More recently, an added dimension has been the availability of porous microcarriers (to replace the solid spheres in which only the surface was utilized) which increases the available surface area a further 10- to 100-fold. Although many of these were initially developed for fluidized beds (e.g. Verax), more recently some have been optimized for stirred suspension.

The key criterion in developing microcarrier technology has been to get the surface chemically and electrostatically suitable for cell attachment, spreading and growth. The initial developmental work by van Wezel and Pharmacia resulted in the Cytodex range of microcarriers [18] and, to put this development in full perspective, the following facts illustrate the scale of opportunity that this method gives:

(i) 1 g of Cytodex microcarrier has a surface area of 6000 cm^2 and, used at a modest concentration of 2 g/liter, gives 12 000 cm^2/liter. This is equivalent to eight large or 15 small roller bottles.

(ii) Suspension culture systems, unlike bottles, etc., can be environmentally controlled, and were optimized in the late 1960s to a minimum 100 liter capacity (i.e. a 1 × 100 liter fermenter was equivalent to 800–1500 roller bottles). The scale-up potential is 10 000 liters (for suspension cells).

Scale-up in cell density can also be achieved by the use of spin filters [19] which allows continuous perfusion of the culture and thus operation at microcarrier concentrations of 15 g/liter and cell densities over 10^7/ml.

Given the statement that successful culture is dependent upon critical experimental techniques, the following general principles are helpful tips:

(i) Cells differ in their attachment requirements, and thus a range of microcarriers should be assessed for suitability (see *Table 7*).
(ii) It is critical to have suitable culture equipment. Spinner flasks should have a special impeller (e.g. Bellco, Wheaton, Techne flasks specifically designed for microcarriers) as a magnetic stirring bar is unsuitable. The shape of the vessel is important for homogeneous mixing, as is the stirring unit, which must operate smoothly without vibration at low revolutions (15–100 rev/min). Culture vessels must be siliconized to prevent microcarriers sticking to the glass, as well as the bottles used for preparing and storing microcarriers.
(iii) Microcarriers can be purchased as dry powders or sterilized solutions. The dry powders are swollen in Ca^{2+}/Mg^{2+}-free phosphate-buffered saline (PBS) (3 h), the supernatant is decanted and discarded, then the residual powder is washed and sterilized in the same buffer (50 ml/g) by autoclaving (103

Table 7. Examples of microcarriers

	Manufacturer	Type	cm^2/g	SG
(a) *Solid*				
Biosilon	Nunc	Polystyrene	255	1.05
Bioglas	Solohill	Glass/latex	350	1.03
Bioplas	Engineering	Polystyrene	350	1.04
		Dextran	6000	1.03
		Collagen	4600	1.04
Cytodex 1, 2	Pharmacia Biotech	Dextran	7000	1.05
Cytodex 3	Pharmacia Biotech	Polyacrylamide	5000	1.03
Dormacell	Pfeifer & Langen		250	1.03
Micarcel G	Reactifs IBF	Dextran	4000	1.03
Microdex	Dextran Products	Gelatin		
Ventegrel	Ventrex Labs			

	Manufacturer	Type	Diameter (μm)	Pore diameter (μm)
(b) *Porous*				
Cellsnow	Kirin	Cellulose	800	>100
Cytoline	Pharmacia Biotech	Polyethylene	1200	>50
Cytopore	Pharmacia Biotech	Cellulose	250	30
Immobosil	Ashby Scientific	Silicone rubber	1000	100

kPa or 1 bar, 15 min, 115°C). *Warning*: do not overheat. The powder microcarriers are 'softer' than glass and plastic and can be used in higher concentrations, thus giving vastly greater unit surface areas (e.g. Cytodex 6000 cm^2/g, glass/polystyrene 300 cm^2/g). Some microcarriers are available at different specific gravities (usually between 1.01 and 1.05), offering a choice depending upon the stirring (or mixing) system.

(iv) Consideration may have to be given to using a supplemented medium during the initial stages of culture to aid cell attachment and to offset the effects of low cell density, which will be more critical in microcarrier than stationary culture. The supplementation may be simple [e.g. nonessential amino acids, pyruvate (0.1 mg/ml), adenine (10 μg/ml), hypoxanthine (3 μg/ml) and thymidine (10 μg/ml)]. Other supplements include tryptose phosphate broth (1 mg/ml), HEPES (5 mM), transferrin (10 mg/liter) and fibronectin (2 μg/ml). Serum, unless serum factors are added, may have to be used at 5–10% initially before being reduced after 1–2 days of culture.

(v) The conventional cell-counting procedure of trypsinizing cells, and then counting in a hemocytometer with Trypan blue, can be tedious or even inaccurate (due to problems in removing cells from the microcarrier sludge). A better method is to use the nuclei-counting method developed by Sanford [20]. The microcarriers are suspended in 0.1 M citric acid containing 0.1%

Crystal violet to lyse the cells and stain the nuclei. The mixture can be vortex mixed or left at 37°C for at least 1 h. An advantage is that samples can be stored for long periods (at least 1 week) at 4°C before being counted. However, it gives a total, not viable, cell count. Cells can also be fixed and stained on microcarriers for microscopic examination using standard procedures. Hematoxylin is the most widely used stain.

(vi) It is possible to re-use some microcarriers using good washing procedures and re-equilibration in PBS. This is not recommended except for glass microcarriers, especially for more than one re-cycle, as performance drops significantly. There are reports of *in situ* re-colonization of microcarriers [21], which is feasible but again reduces the efficiency of the process.

(vii) Enzymic harvesting (e.g. trypsin) of some cells from microcarriers can be difficult or damaging. Consideration can be given to digesting some microcarriers (e.g. by dextranase or collagenase, as appropriate) as this is a far quicker and less damaging means of getting a single-cell suspension.

(viii) Microcarrier culture requires more critical preparation than nondynamic culture systems [10,22]. To ensure success it is very important that all experimental details are carried out optimally. Of particular importance is the quality of the cell inoculum. This should be rapidly dividing, not stationary, and cells should be in good condition (i.e. not trypsin-damaged) and as near a single-cell suspension as possible. It is good practice to feed the seed culture 24 h before use. Also ensure that medium is prewarmed and pre-equilibrated for pH (shifts in pH during attachment are extremely damaging). Due to the number of manipulations in the process, extreme care should be taken with aseptic technique, and as many of the steps as possible carried out in a tissue culture cabinet. Do not inoculate below the minimum number; the culture may not initiate or, if it does, will not reach maximum cell density.

Protocol 27. **Roller bottle culture.** B. Griffiths

Reagents

$5\% \ CO_2/95\%$ air
Standard tissue culture media
Trypsin

Equipment

Inverted microscope
23×12 cm Plastic roller bottles bottle (1400 cm^2; e.g. Becton Dickinson, Sterilin)
Roller bottle apparatus (5–60 rev/h; e.g. Wheaton, Bellco, New Brunswick).

Procedure

1 To a roller bottle, add:
- 1 ml complete culture medium per 5 cm^2 culture area at 37°C
- 5% CO_2 in air to head space [1]
- 1×10^5–2×10^5 cells/cm^2

2 Seal the bottle and place in roller culture apparatus. Rotate the bottle at 12–24 rev/h for the initial attachment phase (2–8 h depending upon cell type). ①

3 Reduce the revolution rate to 5–10 per hour when the culture is growing. Cell growth can be monitored initially under an inverted microscope and

Notes

This procedure takes approximately 5–6 days.

① This faster speed results in an even distribution of cells but should be reduced for cells with low attachment efficiency.
② This can also be carried out to change to a production medium, or when infecting cells, and a lower medium volume can be added to obtain a higher product concentration.
③ Cell yields will be similar to or up to two-fold higher than in stationary cultures, and multilayering (nondiploid cell lines) will occur.

later in the culture period by visual inspection. A medium change can be carried out after 4–5 days if the pH becomes acidic and/or maximum cell densities are required. ②

4 Harvest cells when confluent (after 5–6 days) by removing the medium and trypsinizing in the conventional way (*Protocol 8*). After adding trypsin to a final concentration of 0.25% roll the bottles at speeds of 20–60 rev/h until the cells detach. ③

Pause point

1 Can be left for 24 h to equilibrate before cells are added.

Protocol 28. **Spinner flask culture.** B. Griffiths

Reagents

5% CO_2/95% air
Growth medium
Medium supplements
0.1% Pluronic F-68 (polyglycol) (BASF) ①
Standard tissue culture media ②
0.4% Trypan blue in phosphate-buffered saline ⚠

Equipment

Hemocytometer
Inverted microscope
pH meter
1 liter Spinner flask and magnetic stirrer assembly
Tissue culture cabinet
Water bath at 37°C

Procedure

1 Add 200 ml growth medium to a 1 liter spinner flask, gas with 5% CO_2 and warm to 37°C. [1]

2 Add cells from a logarithmically growing seed culture at 1×10^5–2×10^5 cells/ml. Place the spinner flask on the stirrer and stir at 100–250 rev/min. ③

3 Monitor cell growth at least daily by taking a small sample from the side arm (remove flask to a tissue culture cabinet) and carrying out a cell count using a hemocytometer after Trypan blue staining (*Protocols 13* and *14*).

4 Monitor the pH. In closed systems (i.e. all ports closed with no filters) the pH will become acidic and will need adjusting by day 2–3. ④

Notes

This procedure takes approximately 3–5 days but can be extended.

① To protect cells against mechanical damage, especially at reduced serum concentrations. Carboxymethyl cellulose (15–20 cP) can also be used to protect cells from mechanical damage at 0.1% concentration. Antifoam (6 p.p.m.) (Dow Chemical Co.) is recommended at serum concentrations above 2–3%.

② Can be used with various supplements. Serum at 10% can cause foaming at fast stirring speeds, so the serum level is usually reduced to as low a level as possible (0.5–5%). The use of HEPES buffer to stabilize the pH during the setting-up procedure is beneficial.

5 After 3–5 days the culture density will typically reach 1×10^6–2×10^6/ml (for many hybridomas 0.8×10^6–1.5×10^6/ml only) and can be harvested, or the culture prolonged for extra cell growth with twice-daily 50% medium changes.⑤

③ This is variable depending upon the cell type and geometry of the stirrer bar and vessel: a guideline is to use the minimum speed which gives, by visual examination, complete homogeneity.

④ Remove the vessel to a tissue culture cabinet and either: (i) re-gas with 5% CO_2 in air, (ii) add sodium bicarbonate (5.5%), or (iii) allow cells to settle, remove around 50% of the medium and replace with fresh prewarmed medium.

⑤ Suspension cells tend to have only a limited stationary phase and it is thus important to monitor growth more closely than in monolayer culture to ensure that healthy, rather than dying/dead, cells are harvested.

Pause point

[1] Media may be left to equilibrate for up to 24 h.

Protocol 29. **Microcarrier culture.** B. Griffiths

Reagents

Anchorage-dependent cell line (e.g. MRC-5)
Ca^{2+}/Mg^{2+}-free phosphate-buffered saline (PBS)
5% CO_2/95% air
Eagle's minimum essential medium (MEM; or equivalent alternative) [with 10% fetal bovine serum (FBS), Eagle's nonessential amino acids, and other supplements as necessary]
Microcarrier (e.g. Cytodex 3; Pharmacia)
5% Sodium bicarbonate stock solution
Soybean inhibitor
0.25% Trypsin in PBS

Equipment

37°C Cabinet
Centrifuge
Magnetic stirrer
Microscope
Spinner vessel adapted for microcarrier culture (e.g. Bellco or Techne)

Procedure

1 Add complete medium to the spinner flask (200 ml in 1 liter), gas with 5% CO_2 and allow to equilibrate.

2 Decant the PBS from the sterilized stock solution of Cytodex 3 and replace with growth medium (1 g/30–50 ml). Add Cytodex 3 to spinner vessel to give a final concentration of 2 g/liter (use in the range 1–3 g/liter).

Notes

This procedure takes approximately 4–5 days (minimum).

① Inoculate at the same density per cm^2 as with other culture types (e.g. 5×10^4–$10 \times 10^4/cm^2$). Cytodex 3 at 2 g/liter inoculated at six cells/bead gives:
8×10^6 microcarriers = 4.8×10^7 cells/liter
9500 cm^2 = 5×10^4 cells/cm^2
200 ml = 2.5×10^5 cells/ml

3 Put the spinner on the magnetic stirrer at 37°C and allow the temperature and physiological conditions to equilibrate (minimum 1 h). [1]

4 Add the cell inoculum obtained by trypsinization of late log phase cells (pre-stationary phase) at over five cells/bead (the optimum number to ensure that the number of cells on all beads is seven). ①

5 Place the spinner flask on the magnetic stirrer and stir at the minimum speed to ensure that all cells and carriers are in suspension (usually 20–30 rev/min). It is advantageous if cells and carriers are limited to the lower 60–70% of the culture for this purpose. Alternatively either: (i) inoculate in 50% of the final medium volume and add the rest of the medium after 4–8 h, or (ii) stir intermittently (for 1 min every 20 min) for the first 4–8 h. ②

6 When the cells have attached (expect 90% attachment) the stirring speed can be increased to allow complete homogeneity (40–60 rev/min).

7 Monitor the progress of the culture by taking 1 ml samples at least daily; observe microscopically, and carry out cell (nuclei) counts.

8 As the cell density increases there is often a tendency for microcarriers to begin clumping. This can be avoided by increasing the stirring speed to 75–90 rev/min.

9 After 3–4 days the culture will become acid. Remove the spinner flask and re-gas the head space and/or add sodium bicarbonate (5.5% stock solution). Replace the spinner flask on the stirrer. ③

② Only use this option for cells with very poor plating efficiency as it does cause clumping and uneven distribution of cells per head.

③ With some cells, or at microcarrier densities of 3 g/liter or more, partial medium changes should be carried out. Allow the culture to settle (5–10 min), siphon off at least 50% (usually 70%) of the medium and replace with prewarmed fresh medium (serum can be reduced or omitted at this stage).

④ A cell yield of 1×10^6–2×10^6/ml (2×10^5–4×10^5/cm^2) can be expected.

⑤ As a guide to calculating settled volume and medium entrapment by microcarriers, 1 g of Cytodex, for example, has a volume of 15–18 ml.

10 After 4–5 days the cells reach a maximum cell density (confluency) at the same level as in static cultures, although multilayering is not so prevalent. ④

11 Cells can be harvested in the following way:

- Allow the culture to settle (10 min).
- Decant off as much medium as possible (>90%).
- Add warm Ca^{2+}/Mg^{2+}-free PBS (or EDTA in PBS) and mix.
- Allow the culture to settle and decant off as much PBS as possible.
- Add 0.25% trypsin (30 ml) and stir at 75–100 rev/min for 10–20 min at 37°C.
- Allow the beads to settle out (2 min). Either decant the trypsin plus cells, or pour the mixture through a sterile coarse sinter glass filter or a specially designed filter such as the Cellector (Bellco).
- Centrifuge the cells (800 *g* for 5 min) and resuspend in fresh medium (with serum or trypsin inhibitor, e.g. soybean inhibitor at 0.5 mg/ml). ⑤

Pause point

[1] The vessel can be left for up to 24 h before adding cells.

VI QUALITY ASSESSMENT OF CULTURES. A. Doyle, I. Cour, G. Stacey, T.R. Chen and J. Mowles

Detection of mycoplasma. A. Doyle

The importance of mycoplasma detection in cell cultures should not be underestimated. The concentration of mycoplasmas in the supernatant can be typically in the region of 10^6–10^8 mycoplasmas/ml. Additionally, mycoplasmas will cytoadsorb to the host cells. They do not necessarily manifest themselves in the manner of most bacterial or fungal contaminants (e.g. pH change or culture turbidity). It is important therefore to adopt an active routine detection procedure. Mycoplasmas have been shown to elicit various effects, including the following:

- induction of chromosome aberrations;
- induction of morphological alterations, including cytopathology;
- reduced viability on cryopreservation;
- interference in the rate of growth of cells;
- an influence on nucleic acid and amino acid metabolism;
- induction of membrane alteration and even cell transformation.

Mycoplasmas differ from other prokaryotes in their lack of a cell wall. They are unable to produce even precursors of bacterial cell wall polymers, unlike L forms of bacteria which can do so under the right environmental conditions. Their size is another distinguishing feature; they are the smallest self-replicating prokaryotes, with coccoid forms of only 0.3 μm diameter, capable of reproduction. Their genome size is approximately one-sixth that of *Escherichia coli.*

Mycoplasma contamination is usually caused by any of five common species. The organisms and their natural hosts are *Mycoplasma hyorhinis* (porcine), *M. arginini* (bovine), *M. orale* (human), *Acholeplasma laidlawii* (bovine) and *M. fermentans* (human).

Detection of bacteria and fungi in cell cultures. I. Cour

For any quality control program involving cell lines it is absolutely essential to include routine culture tests for bacteria and fungi. Usually the presence of such contaminants will be obvious even to the novice. This is especially true when the cultures are set up in the absence of all antibiotics; a practice that is strongly recommended. However, while overt microbial contamination may be more common, many organisms will grow slowly in the usual cell cultures, sometimes obscured by animal cell debris, and remain unnoticed.

Identity testing of cell lines. A. Doyle and G. Stacey

The consequences of using a cell line whose identity is not that claimed by the provider are clearly very serious. Failure to identify readily an occurrence of cell line cross-contamination or switching during tissue culture procedure could completely invalidate a body of research. Cell line authentication is therefore an essential requirement for all cell culture laboratories, and should be carried out at both the earliest passages of cultures and at regular intervals thereafter [1]. The occurrence of cross-contamination is not merely anecdotal; documented cases have been widely reported [2,3]. Some earlier reports indicated that levels of cross-contamination may exceed 30% of cultures tested [4]. This figure probably represented an overestimation of the problem; nevertheless, experience at the ECACC in the authentication of cell stocks from a wide range of laboratories indicates that cross-contamination is a neglected problem. The classic example is that of HeLa contamination [2]. A summary of the standard authentication techniques is given below but this list is by no means exhaustive (see also *Table 8*).

Cytogenetic analysis

The microscopic examination of the chromosomal content of a cell line provides a direct method of confirming the species of origin and allows the detection of gross aberrations in chromosome number and/or morphology. Cytogenetic analysis is very

Table 8. Comparison of key features of authentication techniques

	Cytogenetic analysis	Isoenzyme analysis	DNA fingerprinting/ profiling
Determination of species	t	t	—
Rapid identification of a cell line	—	—	t
Detection of a cell line variation	t	—	t
Recognized by regulatory authorities	t	t	*

t, useful applications; *, under consideration.

useful for specific identification of cell lines with unique chromosome markers and has proven useful for the differentiation of cell lines which were apparently identical by isoenzyme analysis. In one study of 47 cell lines reported by O'Brien *et al.* [5], two cell lines could not be differentiated by eight separate enzyme tests, but were readily distinguished by karyology. However, it should be borne in mind that very careful interpretation in the light of considerable experience would be required to differentiate cell lines of normal karyotype beyond the level of species. In addition, many species have not been extensively studied using karyology and therefore the prevalence of chromosome markers in natural populations, and hence in cell lines, may be unknown. Cytogenetic analysis is time-consuming and, to some extent, subjective.

Isoenzyme analysis

Isoenzyme analysis is useful for the speciation of cell lines and for the detection of contamination of one cell line with another. The method utilizes the property that isoenzymes have similar substrate specificity, but different molecular structures. This affects their electrophoretic mobility. Thus, each species will have a characteristic mobility of isoenzymes. While the species of origin of a cell line can usually be determined with only two isoenzyme tests (lactate dehydrogenase and glucose-6-phosphate dehydrogenase), specific identification of a cell line would require a larger battery of tests [4,5].

However, in analysis of enzyme systems with multi-allelic products, low-level expression of some components may lead to unreliable detection and hence mis-identification. In conclusion, the use of a wide range of isoenzyme tests for accurate identification of cell lines requires a detailed knowledge of each isoenzyme system in each species to be studied.

DNA fingerprinting and DNA profiling

Multilocus DNA probes, such as those of Jefferys *et al.* [7], oligonucleotides homologous to very simple sequence repeats [8] and the DNA sequence of the M13 protein 111 gene [9] interact with a widely dispersed range of loci throughout the genomes of many organisms and produce unique fingerprints in Southern blot analysis of cell line DNA. An alternative approach is to analyze single loci bearing variable numbers of tandem repeats (VNTRs) which produce simple DNA profiles (as opposed to fingerprints) comprising one or two bands. The key advantages of this technique in identification are firstly, the apparent simplicity of scoring such profiles, and secondly, that methods based on the polymerase chain reaction (PCR) are available [10,11]. Probes for single loci are used under highly specific conditions, and consequently each one may be useful in only a single species, unlike multilocus fingerprinting. In visualizing many loci, DNA fingerprinting techniques enable recognition of changes in a cell line which might go unnoticed when using single-locus methods. Multilocus DNA fingerprinting can therefore identify cell line variation [12] and cross-contamination [3], and can be used to confirm stability and consistent quality in cell stocks from diverse species. Furthermore, the development of a library of fingerprints from original documented material may help in the validation of suspect cell banks by comparison with the definitive version.

Cell identification by karyology. T.R. Chen

Adequately prepared chromosome slides are essential for successful and meaningful karyologic study. Data collected from these slides are used to delineate essential banding patterns to define the genomic chromosome fingerprint characteristics (GCFC). The chromosome complement of most aneuploid cell lines consists of normal (N) and structurally altered marker (M) chromosomes. The number of copies (NC) of normal (N pattern), NC and kinds of marker chromosomes (M pattern) and polymorphic bands (P pattern) of some chromosomes are unique to each cell population. The P pattern is unique to each individual and exhibits simple Mendelian segregation. It can therefore be used [13] as a genetic signature to identify individual cell lines and clones between the normal diploidies or those even from the same genetic origin.

Laboratory procedures used for assessing the GCFC of a cell population are described. These include air-dried slide preparation (for monolayer, suspension and peripheral blood cultures; *Protocol 34*), conventional nonband staining in 1% Giemsa solution for 10–20 min and four basic chromosome band stainings [i.e. trypsin–Giemsa (G), quinacrine-fluorescent (Q), constitutive-heterochromatin (C) and reverse-Giemsa (R); *Protocol 35*]. The four band staining procedures are chosen since they represent the basic and most broadly used methods for characterization of a cell population by karyology.

G/R bands

The G or R banding is generally used for both gross and detailed analyses of the chromosome complement of a cell. If adequately prepared, these bandings give sufficient and precise information on most individual chromosomes in a cell complement, including the identification of segments and sites involving inter-chromosome exchanges (*Figure 7a* and *b*). Both methods produce equally informative banding patterns. Chromosome spreads, with a moderate degree of high-resolution sub-bands, are found in slides prepared by procedures described in *Protocol 34* (i.e. the resolution is at a range of 500–600 sub-band level in a normal human chromosome complement). Likewise, some polymorphic properties of chromosomes can also be delineated.

Q bands

Q banding gives a pattern similar to G bands (*Figure 7c*). It delineates clearly the polymorphic property of some chromosomes, and is simple to process. It produces recognizable banding patterns even from a moderate-quality slide preparation and has little effect on gross chromosome morphology. The Q-banded metaphases can be re-stained to reveal different bandings. Q banding is routinely used for locating intact and/or translocated human Y chromosomes.

C bands

C-band staining identifies the constitutive heterochromatin that resides at the specific site of a chromosome, often at the pericentric region (*Figure 7d*). Since it exhibits discrete and readily recognizable staining properties, it can be used to identify some chromosome pairs. The amount and location of constitutive heterochromatin may differ between chromosomes and show the trait of simple Mendelian [14] inheritance. Thus the C-band polymorphism can be used for distinguishing individual traits in a family [15]. Distinct C band pattern are also noted between some species. In man/mouse interspecies somatic cell hybrids, for example, C bands can be used for distinguishing between human and mouse chromosomes [16], as well as hybrid chromosomes consisting of segments from both species. Inter-chromosome exchanges are often seen in aneuploid cells derived from cancer tissues, for example. These exchanges frequently occur at or near the pericentric region of a chromosome, where the C band resides in most chromosomes. Thus this is a convenient marker for providing supplementary evidence for identifying the involvement of a specific chromosome.

Elimination of contamination. J. Mowles

In the event of cell cultures becoming contaminated with bacteria, fungi or mycoplasma, the best course of action is to discard the culture, check cell culture reagents for contamination, thoroughly disinfect all safety cabinets and work surfaces, and resuscitate a fresh culture from previously frozen stock. In the case of contamination with a spore-forming organism,

---

Figure 7. (a) Giemsa-stained G banding from a human colorectal carcinoma cell line. Chromosomes were stained by the Fast trypsin–Giemsa method. There are many structurally altered marker chromosomes resulting from deletions and inter-chromosomal translocations. (b) Giemsa-stained R banding. This is from another human colorectal carcinoma line, and from a slide after BrdU/TdR treatment. (c) Quinacrine mustard-stained Q banding from a normal human female metaphase. The long arrow indicates the brightly fluorescent polymorphic Q band of chromosome N3 at the centromere, and the short arrow, that of chromosome N21 at the terminal short-arm. (d) Constitutive heterochromatic C banding. This is an endo-reduplicate metaphase from a man–mouse somatic cell hybrid. All but one mouse chromosome had darkly stained centromeric, and occasionally also interstitial, C bands. Arrow indicates two human chromosomes that remained in this hybrid line.

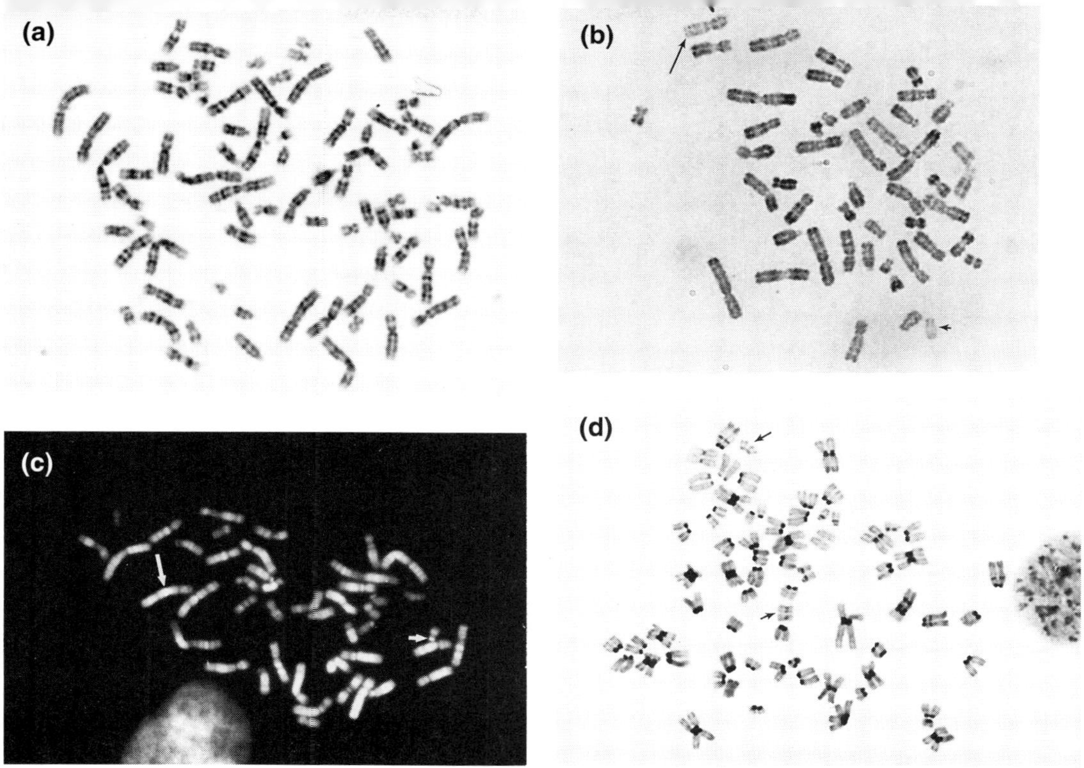
(a)
(b)
(c)
(d)

and where such facilities exist, room fumigation may also be advisable. However, in the case of irreplaceable stocks this course of action may not be possible and antibiotic treatment may be necessary to eliminate the contamination. There is by no means any guarantee of success and the antibiotic has to be chosen with care, based on some knowledge of the contaminating organism(s).

For most laboratories, antibiotic treatment is the most convenient method of eradication, however, with mycoplasma in particular, there are alternatives available. These include passage in athymic nude nice [17], growth of cells in rabbit or guinea pig serum [18] and the use of nucleic acid analogs [19]. They have not become methods of first choice and should be viewed as last resort treatments.

Methods available

Detection of mycoplasma (*Protocols 30* and *31*)
A range of assay techniques are available for the detection of mycoplasma infection. These include staining, culture, DNA probes and PCR. To remove the risk of false positives and false negatives, at least two methods should be employed.

Procedures for two routinely used methods are given: DNA stain (*Protocol 30*) and microbiological culture (*Protocol 31*). Hoechst DNA stain has the advantage of relative rapidity – a result can be obtained within 48 h – whereas broth and agar culture can take up to a month to complete. On the other hand, DNA stain lacks sensitivity, detecting 10^5 colony-forming units (cfu)/ml as compared with a

References

1. Hay, R.J. (1988) *Analyt. Biochem.* **171**, 225.
2. Nelson-Rees, W.A., Daniels, D.W. and Flandermyer, R.R. (1981) *Science* **212**, 446.
3. Van Helden, P.D., Wild, I.J.F., Albrecht, C.F., Theron. E., Thornley, A. L. and Hoal-van Helden, E.G. (1988) *Cancer Res.* **48**, 5060.
4. Halton, D.M., Peterson, W.D. and Hukku, B. (1983) *In Vitro* **19**, 16.
5. O'Brien, S.J., Shannon, J.E. and Gail, M.H. (1980) *In Vitro* **16**, 119.
6. Wright, G.C., Daniels, W.P. and Fogh, J. (1981) *J. Natl Cancer Inst.* **66**, 2349.
7. Jefferys, A.J., Wilson, V. and Thein, S.L. (1985) *Nature* **316**, 76.

theoretical 1 cfu sample with the culture method. Both these methods have the added benefit of regulatory approval [20].

With both methods the cells to be tested should, before testing, complete at least two passages in antibiotic-free media, as infection may be hidden by the presence of antibiotics. Cell cultures tested from frozen ampoules should undergo at least two passages because cryoprotectants may also mask infection [21,22].

Detection of bacteria and fungi (*Protocols 32* and *33*)
Two isolation media are suggested by the US Code of Federal Regulations and the European Pharmacopoeia as being suitable for revealing the presence of viable forms of bacteria, fungi and yeasts. These are fluid thioglycollate medium (FTM) and soybean casein digest (tryptone soya broth, TSB). The ECACC 'in-house' detection method is described in *Protocol 32,* and an alternative method is described in *Protocol 33*.

Antibiotic elimination of contaminants
Cells may be cultured in the presence of the chosen antibiotic (see *Table 9*) for 10–14 days in order to eliminate contamination. Each passage should be performed at the highest dilution of cells at which growth occurs. If the contaminant is still detectable after this time, it is unlikely that the antibiotic used will prove successful, and another antibiotic should be tried.

8. Ali, S., Muller, C.R. and Epplen, J.T. (1986) *Hum. Genet.* **74**, 239.
9. Ryskov, A.P, Jincharadze, A.G., Prosnvak, A.G., Ivanov, P.L. and Limborskaya, K.(1988) *FEBS Lett.* **133,** 388.
10. Jefferys, A.J., McLeod, A., Tamaki, K., Neil, D.L. and Monkton, D.G. (1991) *Nature* **354**, 204.
11. Tautz, D. (1989) *Nucleic Acids Res.* **17**, 6469.
12. Thacker, J., Webb, M.B.T. and Debenham, P.G. (1988) *Som. Cell Mol. Genet.* **14**, 519.
13. Chen, T.R., Dorotinsky, C.S., McGuire, L.J., Macy, M.L. and Hay, R.J. (1995) *Cancer Genet. Cytogenet.* **81**, 103.
14. Craig-Holmes, A.P. (1977) In *Population Cytogenetics* (E.B. Hook and I.H. Porter, eds), p. 179. Academic Press, New York.
15. Chen, T.R. and Ruddle, F.H. (1974) *Proc. Soc. Exp. Biol. Med.* **147**, 533.
16. Boone, C., Chen, T.R. and Ruddle, F.H. (1972) *Proc. Natl Acad. Sci. USA* **69**, 510.
17. Van Diggelen, O.P., Shin, S. and Phillips, D.M. (1977) *Cancer Res.* **37,** 2680.
18. Nair, C.N. (1985) *Proc. Soc. Exp. Biol. Med.* **179,** 254.
19. Marcus, M., Lavi, U., Nattenberg, A., Rottem, S. and Markowitz, D. (1980) *Nature* **285**, 659.
20. US FDA/CBER (1993) *Points to Consider.*
21. Doyle, A. and Bolton, B.J. (1994) in *Basic Cell Culture* (J. Davis, ed.), pp. 243–270. IRL Press, Oxford.
22. Mowles, J.M. (1988) *Cytotechnology* **1,** 355.
23. Cour, I., Maxwell, G. and Hay, R.J. (1979) *Tissue Cult.*

During the course of antibiotic treatment, and also 5–7 days after treatment in the case of bacteria and fungi, and 25–30 days after treatment for mycoplasma-contaminated lines, re-test the culture by an appropriate method. Regular routine testing should, however, occur after this period. This is particularly important in the case of mycoplasmas, where low levels of infection may well persist after antibiotic treatment.

In the case of bacteria and fungi, a Gram stain would give an indication of which antibiotic to use (see *Tables 9* and *10*).

Table 9. Antibiotics active against bacteria

Antibiotic	Working concentration (mg/liter)	Gram-positive bacteria	Gram-negative bacteria
Ampicillin	100	✓	✓
Cephalothin	100	✓	✓
Gentamycin	50	✓	✓
Kanamycin	100	✓	✓
Neomycin	50	✓	✓
Penicillin	100	✓	
Polymyxin B	50		✓
Streptomycin	100	✓	✓
Tetracycline	10	✓	✓

Assoc. Man. **5**, 1157.
24. Freshney, R.I. (1987) *Culture of Animal Cells – a Manual of Basic Techniques*, 2nd edn. Alan R. Liss, New York.

Protocols provided

30. *Detection of mycoplasmas using DNA stain*
31. *Microbiological culture*
32. *Detection of bacteria and fungi by isolation*
33. *Detection of bacteria and fungi in cell cultures*
34. *Air-dried chromosome slide preparation*
35. *Staining of G, Q, C and R bands*

Various antibiotics have been investigated for their ability to eradicate mycoplasmas from cell lines. Historically the antibiotics of choice were either minocycline or tiamulin. However, many cell culture mycoplasmas are now resistant to these antibiotics. Today the most effective are the quinalone antibiotics ciprofloxacin [21] and MRA (see *Table 10*).

It is extremely important that after antibiotic treatment the cells are maintained in antibiotic-free media. Lack of evidence of mycoplasma infection, for example, does not necessarily indicate that the culture is free of such infection, as the level of infection may be below the limit of detection.

Table 10. Commercially available antibiotics

Antibiotics active against fungi	
Amphotericin B (Sigma)	2.5 mg/liter
Ketaconazole (Sigma)	10 mg/liter
Mystatin (Sigma)	50 mg/liter
Antibiotics active against mycoplasmas	
Ciprofloxacin	20 mg/liter
MRA (mycoplasma-removal agent; ICN-Flow)	0.5 mg/liter

Protocol 30. **Detection of mycoplasmas using DNA stain.** A. Doyle

Reagents ①

Carnoy's fixative (3 ml methanol and 1 ml glacial acetic acid, prepared fresh) ⚠
100 ml Hoechst stain stock solution (Bisbenzimide-Hoechst 33258) ②⚠
50 ml Hoechst stain working solution (50 μl stock solution in 50 ml water, prepared fresh)⚠
Mountant ③
Swab

Equipment

Coverslips
Fluorescence microscope with epi-fluorescence (e.g. Zeiss) and 340–380 nm excitation filter, and 430 nm suppression filter
Microscope slides
Tissue culture dishes (33 mm)

Procedure

Detection procedure

1 Using a routine method of subculture, harvest adherent cells (e.g. trypsin) and resuspend in the original culture medium to a concentration of about 5×10^5 cells/ml. Test the suspension lines direct from the culture at this concentration.

2 Add 2–3 ml test cells to each of two tissue culture dishes containing glass coverslips. Incubate at 37°C in a humidified 5% CO_2/95% air atmosphere for 12–24 h.④

Notes

This procedure takes approximately 30 minutes. Results are available between 1 and 3 days post-testing.

① Use analytical grade (Analar) reagents and distilled water.
② Add 10 mg Hoechst 33258 to 100 ml distilled water; filter sterilize using a 0.2 μm filter; wrap the container in aluminum foil and store in the dark at 4°C.
③ Autoclave 22.2 ml 0.1 M citric acid with 27.8 ml 0.2 M disodium phosphate; mix with 50 ml glycerol; adjust pH to 5.5; filter sterilize and store at 4°C.

3 Remove one dish and leave the remaining dish for preparation after a further 48 h.

4 Before fixing the cells, examine under an inverted microscope for bacterial and fungal contamination at ×100 magnification.

5 Fix the cells by adding approximately 2 ml Carnoy's fixative dropwise to the edge of the dish. Leave for 3 min at room temperature. It is particularly important for suspension cultures to add fixative in this way as it avoids cells being swept to one side of the culture dish.

6 Decant the fixative to a waste bottle (avoiding fumes) for careful disposal, add a further 2 ml aliquot of fixative to the dish and leave for 3 min at room temperature, then decant the fixative to a waste bottle.

7 Allow the coverslip to air dry. Invert the lid of the dish. Use forceps to rest the coverslip against the lid for about 30 min.

8 Add 2 ml Hoechst stain (wearing gloves) to the coverslip. Leave for 5 min at room temperature. Shield the coverslip from direct light at this point.

9 Decant the stain to a waste bottle.

10 Add one drop of mountant to a labeled slide. Place the coverslip cell side down on a slide.

11 Under ultra-violet epifluorescence examine the slide at ×100 magnification with oil immersion.⑤

④ Experience of working with any particular cell line should remove the absolute necessity for an accurate cell count. An adequate number of cells should be added to dishes so that a semi-confluent spread of cells on the coverslip is obtained at the time of observation (at 1 day and 3 days post-incubation). Cell overgrowth would make interpretation difficult. If it is thought that after 3 days' growth the media would be exhausted, the cells should be resuspended in a mixture of their original medium and fresh culture media.

⑤ The fluorochrome dye Hoechst 33258 binds specifically to DNA. Cultures infected with mycoplasma are seen under fluorescence microscopy as fluorescing nuclei with extranuclear mycoplasmal DNA, whereas uninfected cell cultures contain fluorescing nuclei against a negative background. Mycoplasmas may appear as filamentous forms, some of which may branch, indicating a culture in logarithmic growth, or as cocci, which is typical of an aged mycoplasma culture. Some slide preparations may contain extranuclear fluorescence produced by disintegrating nuclei and these should not be confused with mycoplasmas. Normally, fluorescence of this type is not uniform in size and is too large to be due to mycoplasmas. Contaminating fungi or bacteria will also stain using this technique but will appear much brighter and larger than mycoplasmas. Both positive and negative slides may be kept in the dark for several weeks without deterioration.

Use of indicator cell lines

An alternative to the described procedure is the use of an indicator cell line on to which the test cells are inoculated. The advantages are the standardization of the system and the increased surface area of cytoplasm in relation to nucleus area. This aids in detecting the mycoplasma, and also enables the screening of serum and other cell culture reagents which may be inoculated on to the indicator line. Positive and negative controls may be included in each assay. However, it may prove impracticable for many laboratories to grow mycoplasma cultures to use as positive controls. The disadvantages of using an indicator line include the time and effort required for preparation.

A recommended indicator cell line is the Vero African green monkey kidney cell line which has a high cytoplasm/nucleus ratio. The indicator cells are added at a concentration of approximately 1×10^4 cells/ml to coverslips in tissue culture dishes 10–24 h before inoculation. The test sample should be added at a concentration of approximately 5×10^5 cells/ml to give a semi-confluent monolayer at the time of observation (at 1 and 3 days post-inoculation of the sample). Slides are prepared as in the direct method.

Reagents

80 ml Agar media
Dienes stain (2.5 g Methylene blue, 1.25 g Azure II, 10.0 g maltose, 0.25 g Na_2CO_3, 100 ml distilled water. Dilute this stock solution 1:3 with distilled water)
Horse serum (TCS Biologicals) ①
70 ml Mycoplasma broth base (Oxoid)
10 ml Pig serum ②
Swab
10 ml Yeast extract (Oxoid) ③

Equipment

Autoclave
Glass vials
37°C Incubator
Inverted microscope and slides
Petri dishes
Pipettes and pipettor
Swab

Procedure

Agar preparation

1 Autoclave agar at 103 kPa (15 lb/in^2, 121°C) for 15 min.④ Cool to 50°C and mix with the other constituents (which have been warmed at 50°C).

2 Dispense 8 ml per 5 cm diameter petri dish. Store the dishes at 4°C in sealed plastic bags. Prepared medium must be used within 10 days.

Broth media preparation

1 Add 2 g broth base to 70 ml water. Mix and autoclave at 103 kPa (15 lb/in^2, 121°C) for 15 min.

Notes

This procedure takes approximately 1 hour. Results are available 4 weeks after testing.

① Dispense 20 ml aliquots in universal containers. Do not heat inactivate. Store at −20°C.
② Dispense in universal containers in 10 ml aliquots. Heat to 56°C for 45 min. Store at −20°C.
③ Add 7 g yeast extract to 100 ml water, mix and autoclave at 103 kPa (15 lb/in^2, 121°C) for 15 min. Dispense in universal containers in 10 ml aliquots. Store at 4°C.

2 Mix the constituents, dispense in glass vials in 1.8 ml aliquots and store at 40°C. Complete medium may be stored without deterioration for several weeks.

Cell culture

1 Using a sterile swab, harvest adherent cells. Resuspend them in the culture medium to a concentration of about 5×10^5 cells/ml. Test the suspension cells direct from the culture at this concentration.

2 Inoculate an agar plate with 0.1 ml of the test sample and the broth with 0.2 ml of the test sample. Incubate the test plate under anaerobic conditions (Gaspak, Becton-Dickinson Microbiology Systems) at 37°C for 21 days, and the test broth aerobically at 37°C.

3 At approximately 7 and 14 days post-inoculation, subculture test broth on to agar plates, and incubate anaerobically at 37°C.

4 After 7, 14 and 21 days incubation, agar plates are examined under ×40 or ×100 magnification using an inverted microscope.

Culture of mycoplasma

1 Each new batch of media ingredients should be subject to quality control before agar or broth preparation. It is especially important to show that new batches of pig and horse serum can support the growth of a representative sample of species found infecting cell cultures (e.g. any two of *Mycoplasma orale*, *M. hominis*, *M. fermentans*, *M. arginini*, *M. hyorhinis* or *Acholeplasma laidlawii*). The National Collection of Type

④ Add 2.8 g mycoplasma agar (Oxoid) to 80 ml water; mix and autoclave at 1–3 kPa for 15 min. Prepare fresh as required.

Cultures or the American Type Culture Collection may supply type strains, or wild-type strains may be used. Stock positive control cultures may be kept frozen at −70°C in mycoplasma broth.

2 Quality control of each batch of complete broth media should be performed before use by adding at least two different strains of positive control mycoplasmas. A noninoculated broth should be incubated as a negative control.

3 Mycoplasma agar medium should be shown to be able to support mycoplasma growth. The test can be performed at the time of the test sample inoculation. A noninoculated plate should also be included as a negative control.

4 It is necessary to be able to distinguish 'pseudocolonies' and cell aggregates from mycoplasma colonies on agar. Pseudocolonies are caused by crystal formation and may possibly increase in size. They are distinguished from genuine mycoplasma colonies by using Dienes stain, which does not stain pseudocolonies but stains mycoplasma colonies blue. Most fungal and bacterial colonies also appear colorless. Cell aggregates, however, do not increase in size and thus are more easily distinguished. An indicator of genuine mycoplasma colonies is their typical 'fried egg' appearance on agar. This, however, is not always apparent in primary isolates. By using a sterile bacteriological loop, cells may be disrupted, leaving the agar surface free of aggregates, but mycoplasma colonies, because of the nature of their growth, will leave a central core embedded in the agar.

Protocol 32. **Detection of bacteria and fungi by isolation.** I. Cour

Reagents

Fluid thioglycollate medium (FTM)
Soybean casein digest (tryptone soya broth, TSB)
Positive controls (*Bacillus subtilis*, ATCC 6633; *Candida albicans*, ATCC 11437; *Clostridium sporogenes*, ATCC 10231)

Equipment

22°C and 32°C Incubators
Pipettes and pipettors

Procedure

1 Prepare seven batches of FTM and six batches of TSB and inoculate them as follows:

- inoculate two FTM and two TSB broths with test sample and incubate one broth of each pair at 32 ± 1°C and 22 ± 1°C for 14 days.
- Inoculate two FTM and two TSB broths with *B. subtilis* and *C. albicans* and incubate one broth of each pair at 32 ± 1°C and 22 ± 1°C for 14 days.
- Inoculate one FTM broth with *C. sporogenes* and incubate at 32 ± 1°C for 14 days.
- Two FTM and two TSB broths will remain uninoculated (negative control). Incubate one broth of each pair at 32 ± 1°C and 22 ± 1°C for 14 days.

Notes

This procedure takes approximately 30 min plus a 14 day incubation period.

2 Observe the broths after 14 days. The presence of bacteria, fungi or yeast will be indicated by turbidity in the broth media. The test is valid if all positive-control broths show evidence of bacteria, fungi or yeast and the negative-control broths show no evidence of bacteria, fungi or yeast.

Protocol 33. **Detection of bacteria and fungi in cell cultures.** I. Cour

Reagents

Cell cultures ①
Positive controls (*Bacillus subtilis*, ATCC 6633; *Candida albicans*, ATCC 11437; *Clostridium sporogenes*, ATCC 10231)

Equipment

GasPak Anaerobic System
Hot plates and stirring apparatuses
Inverted microscope with phase contrast optics
Pipettes and pipettors

Procedure ②

Microscopic examination ③

1 Using an inverted microscope, equipped with phase contrast optics if possible, examine cell culture vessels individually. Scrutiny should be especially rigorous in cases in which large-scale production is involved. Check each culture first using low power.

2 After moving the cultures to a suitable isolated area, remove aliquots of fluid from cultures that are suspect and retain these for further examination. Alternatively, autoclave and discard all such cultures.

3 Prepare wet mounts using drops of the test fluids and observe under high power.

Notes

① Suitable media would be bacto blood–agar base (Difco), bacto brain–heart infusion agar (Difco), bacto Sabouraud dextrose broth (Difco), bacto thioglycollate medium (Difco), bacto yeast extract (Difco), dehydrated media for bacterial/fungal cultivation, nutrient broth base (Difco), trypticase soy broth powder or YM broth base (Difco). The suppliers listed provide the specific media required but other suitable vendors exist. Batches of media should be tested for optimal growth promotion before use in cell culture quality control.

② While this procedure will permit detection of most common bacterial and fungal organisms that grow in cell cultures, it has been noted that at least one, very

4 Prepare smears, heat-fix and stain by any conventional method (e.g. Wright's stain), and examine under oil immersion.

5 Consult refs 23 and 24 for photomicrographs of representative contaminants and further details.

Culture tests

1 Pool and mix the contents of about 5% of the ampoules from each freeze lot prepared using narrow-width pipettes. Aliquots from cultures to be tested should generally include some of the monolayer or cell suspension. It is commonly recommended that antibiotics be omitted from media used to cultivate and preserve stock cell populations. ④

2 From each pool, inoculate each of the test media listed in *Table 11* with a minimum of 0.3 ml of the test cell suspension and incubate under the conditions indicated. Include positive and negative controls comprising a suitable range of bacteria and fungi which might be anticipated. A recommended grouping consists of *Pseudomonas aeruginosa*, *Micrococcus salivarius*, *Escherichia coli*, *Bacteroides distasonis*, *Penicillium notatum*, *Aspergillus niger* and *Candida albicans*.

3 Observe as suggested for 14–21 days before making a conclusion as to whether the test is negative or not. Contamination is indicated if colonies appear on solid media or if any of the liquid media become turbid.

fastidious, bacterial strain initially escaped observation. This was present in nine different cultures from a single clinical laboratory in the USA submitted for testing and expansion under a government contract. The organism grew extremely slowly but could be detected after 3 weeks incubation with cell cultures that had no antibiotics and had no fluid changes. Samples so developed were inoculated to sheep blood agar plates and New York City broth [ATCC (American Tissue Culture Collection) medium 1685]. The organism could be observed during a subsequent 6 week incubation period at 30°C.

The Bacteriology Department at ATCC determined the appropriate culture conditions for this microorganism, and tentatively identified it as a *Corynebacterium*. Antibiotic sensitivity tests revealed bacteriostasis with some compounds but no bacteriocidal antibiotics have yet been found.

This incident emphasizes the critical importance of diligent testing of cell cultures for contaminant microorganisms. By combining procedures such as those described here, with procedures included elsewhere in this volume (e.g. fluorescent or nucleic acid probes for mycoplasma and viruses) one can be more certain that clean cell cultures are available for experimentation.

③ Microscopic examination is only sufficient for detection of gross contaminations, and even some of these cannot be readily detected by simple observations. Therefore, an extensive series of culture tests (see following instructions)

Table 11. Regimen for detecting bacterial or fungal contamination in cell cultures

Test medium	Temperature (°C)	Aerobic state	Observation time (days)
Blood agar with:			
fresh defibrinated	37	Aerobic	14
rabbit blood (5%)	37	Anaerobic	14
Thioglycollate broth	37	Aerobic	14
	26		
Trypticase soy broth	37	Aerobic	14
	26		
Brain–heart infusion broth	37	Aerobic	14
	26		
Sabouraud broth	37	Aerobic	21
	26		
YM broth	37	Aerobic	21
	26		
Nutrient broth with 2% yeast	37	Aerobic	21
	26		

For further details see ref. 23

is also required to provide reasonable assurance that a cell line stock or medium is free of fungi and bacteria.

④ If antibiotics are used, the pooled suspension should be centrifuged at 2000 *g* for 20 min and the pellet resuspended in antibiotic-free medium. A series of three such washes with antibiotic-free medium prior to testing will reduce the concentration of antibiotics which could obscure contamination.

Protocol 34. **Air-dried chromosome slide preparation.** T. R. Chen

Reagents ①

Anhydrous methanol (keep at 40°C)⚠
100 μg/ml Colcemid ⚠
Glacial acetic acid ⚠
0.075 M KCl (hypotonic; hypo-KCl)
Hypotonic TV [hypo-TV; 1 g glucose, 2 g NaCl, 0.2 g KCl, 0.2 g disodium-EDTA, 2.18 g $Na_2HPO_4.7H_2O$, 0.2 g KH_2PO_4, 1 g trypsin (Difco; 1:250), 3 ml Phenol red (0.5%, w/v) and 1000 ml distilled water. Adjust pH to 7.2–7.4. Keep frozen and use at 37°C]

Equipment

Automatic pipettor
15 ml and 50 ml Centrifuge tubes plus rack
Crushed ice
Disposable Pasteur pipettes and rubber bulbs
Microscope slides with the frosted edge. (Re-clean with a detergent, soak in distilled water and keep at 40°C)
Bench-top centrifuge
Vacuum device to discard liquid waste products.

Procedure (see *Figure 8*) ②

1 Plate 2×10^6–3×10^6 cells in 25 ml growth medium per T-75. Leave for 24 h.

2 Add colcemid to a final concentration of 0.05–0.1 μg/ml and incubate at 37°C for 30 min (fast-growing cells) to over 3 h (slow-growing cells).

3 Decant the medium and add 6 ml hypo-TV. Incubate for 10 min at 37°C. ① [1]

Notes

This procedure takes approximately 2–2.5 hours following the 24 h growth period.

① All solutions should be filter-sterilized.
② This procedure is for use with monolayer cultures. The air-dry preparation method for suspension and peripheral blood cultures is the same as that described here except that the hypotonic TV step (step 3) is replaced by KCl treatment.

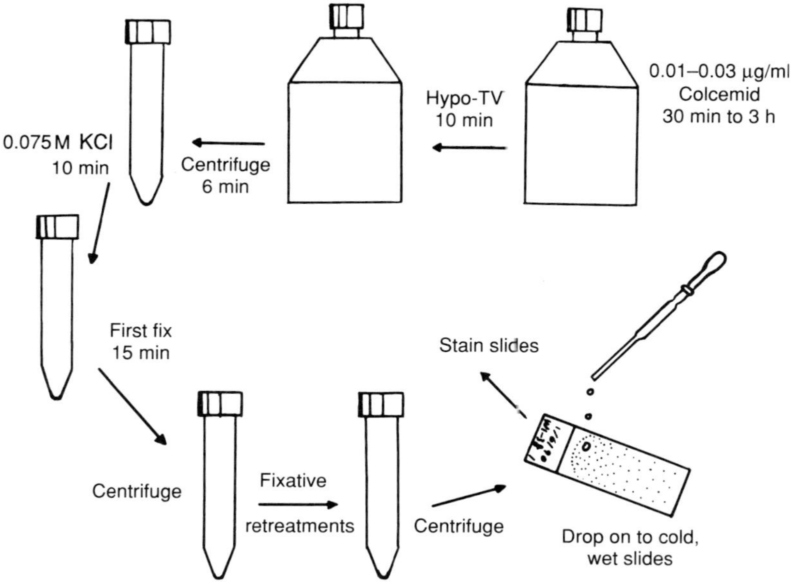

Figure 8. Stepwise procedure for slide preparation.

③ Keep the volume of the cell pellet to less than 0.25 ml per 15 ml centrifuge tube.

④ Always try to resuspend cells to form a uniform cell suspension before adding fresh fixative.

⑤ For a blood culture in the first fixation, resuspend cells in 10–15 times the volume of supernatant and add a pipetteful of fixative each time to a final total volume of 10–15 ml. Thereafter follow the same fixation procedure as described.

⑥ Use glacial acetic acid and 100% methanol. Do not use old solutions in bottles that were opened more than 1 month previously. Methanol should be kept under refrigeration after the bottle is opened.

⑦ The optimum atmospheric moisture conditions for the slide preparation are 30–40% humidity. Wear a laboratory coat and gloves, and have an autoclavable bag on hand for collecting waste materials.

4 Collect the cells into a 15 ml centrifuge tube prefilled with 0.6 ml serum (for enzyme inactivation). Mix and centrifuge at a standard condition (100–170 *g* for 6 min). ③

5 Discard the supernatant, slowly re-suspend cells in KCl (add dropwise, with continuous agitation at first, and gradually increase volume) to a total of 3 ml. Leave at room temperature for 10 min. ④⑤

6 Spin down, discard most of the supernatant but leave a volume equal to the cell pellet for resuspending cells.

7 Add freshly mixed, cold 1:3 acetic acid/methanol fixative, in the same manner as for KCl, to about 3.5 ml total cell suspension, and spin down after 15 min. ⑥

8 Decant the supernatant totally. Repeat the fixation step and centrifuge after 15 min.

9 Repeat the fixation step after 10 min and spin down.

10 Discard most of the supernatant fixative but leave 10–15 times the volume of the cell pellet to resuspend the cells.

11 Pick up a chilled, clean and wet slide with a pair of forceps, hold it at the frosted edge with fingers, shake off excess water, hold at a slightly downward angle, place a few drops of cell suspension from its frosted edge allowing to run down toward the other edge and promptly blow this gently to spread cells evenly over the slide surface. ⑦

12 Drain off excess solution on the slide by making a fanning motion a few times and air-dry at room temperature overnight before staining. Store slides not to be used immediately for staining in a tightly sealed desiccator at 4°C in a slide box.

Pause point

[1] The fixation time may be extended to 1 h.

Protocol 35. **Staining of G, Q, C and R bands.** T. R. Chen

Reagents

2× SSC (1:1 mix of 0.6 M NaCl and 0.06 M sodium citrate, pH 7)
Barium hydroxide (saturated solution in distilled water) (Sigma)
1× Earle's saline solution
G buffer (0.025 M monobasic potassium phosphate. Adjust to pH 6.8–7.0 with 10% NaOH)
Giemsa stain ①⚠
Giemsa stain mixture ②⚠
0.2 M and 1 M HCl ⚠
Q buffer (27.8 ml 0.2 M Na_2HPO_4 and 22.2 ml 0.1 M citric acid, pH 5.5)
Quinacrine mustard (QM) stain [0.1% (w/v) in distilled water] ⚠③

Equipment

Centrifuge tubes
50 ml Coplin jars
No. 1 Coverslips, 22 × 50 cm^2
Pipettes, pipettor and micropipettors
Suctioning apparatus

Procedures

Fast trypsin–Giemsa G banding

1 Incubate 1- to 3-day-old slides in a dry oven at 60°C for 4–6 h.

2 Soak these slides in G buffer at 60°C for 11 min.

3 Distribute 0.95 ml of trypsin/stain mixture evenly over the slide. Incubate at 4°C for 15 min, rinse thoroughly with distilled water and air-dry.

QM staining

1 Stain in 0.1% QM stain solution in a Coplin jar in the dark at room temperature for 30–45 min.

Notes

① Dissolve 1 g Giemsa powder in 66 ml glycerol at 60°C for 12 h, then add 66 ml anhydrous methanol and incubate at 60°C for 24 h. Store for 1 month before use.

② Add 0.55 ml trypsin (1%) to a mixture of 6.5 ml G buffer/2.5 ml 100% methanol, add 0.23 ml Giemsa stain, then mix thoroughly but gently.

③ QM is a potent carcinogen. Wear globes and a laboratory coat, and work in a chemical hood when handling it. Follow the guidelines proposed by the Sigma-Aldrich Corporation (1988) for disposing of QM solution.

2 Wash thoroughly in distilled water then mount in Q buffer with a clean coverslip.

C banding

1 Soak in 0.2 M HCl at room temperature for 1 h and rinse thoroughly with distilled water.

2 Incubate in the saturated $Ba(OH)_2$ at room temperature for 1 h then rinse thoroughly.

3 Incubate in 2× SSC at 58–60°C for 1 h.

4 Rinse briefly in G buffer at room temperature and stain in 2–3% Giemsa solution (e.g. 1.5 ml Giemsa stock to 50 ml G buffer and 1.5 ml 100% methanol for 3% Giemsa solution) for 20–30 min. Rinse thoroughly with distilled water and air-dry.

R banding

1 Adjust Earle's buffer saline solution to pH 6.5 with 1 M HCl. Place in a Coplin jar and heat to 88°C in a water bath.

2 Warm the slides briefly and place one slide at a time in the hot buffer for 11 min. Rinse briefly in distilled water then stain for 10 min in 3% Giemsa staining solution. Rinse in distilled water and check the staining condition, then air-dry, if adequately stained. The slide can be re-stained in the original staining solution.

VII IMMORTALIZATION. C. MacDonald and B. Bolton

Immortalization methods: overview. C. MacDonald

Cells isolated from tissues can be used to set up primary cultures which may be maintained *in vitro* for a limited period of time. Some cell types proliferate and cultures can be kept for weeks or months, whereas others do not divide and the cells degenerate rapidly. However, even those cells which are capable of division have only a limited *in vitro* lifespan [1], and as the cells approach the end of this lifespan an increasing proportion of them become unable to synthesize DNA. The cells then enter a crisis phase, cell division ceases and the culture becomes senescent. This chapter describes some of the approaches that have been used to immortalize cells to enable proliferating cultures to be established and maintained over prolonged periods.

Although both immortal and transformed cell lines are capable of prolonged growth the question of immortality or indefinite growth is problematic – how long must a culture last to be considered to be immortal. It may be preferable, therefore, to use the term 'extended lifespan' since this avoids the presumption of indefinite cell division. Transformed cell lines are often considered to be permanent, but SV40-transformed human fibroblasts for example, normally go through a crisis of senescence, the onset of which is delayed in comparison with untransformed cells, rather than abolished [2].

In rodent cells, immortalization appears to occur as a single step, and may be an intermediate stage to transformation. In general, immortal cell lines retain some or all of the differentiated characteristics of the original tissue from which they were isolated. Transformed cells, in contrast, have usually lost their differentiated phenotype, have an abnormal chromosome complement and are tumorigenic. Transformation and immortalization appear to be more complex in human cells compared with rodent cells. Immortalization in human cells appears to occur in two stages, one of which is equivalent to the rodent immortalization step and can be mediated by the same oncogenes [3].

Epstein–Barr virus transformation of B lymphocytes. B. Bolton

Epstein–Barr virus (EBV) is a herpes virus which naturally infects human lymphocytes, causing infectious mononucleosis. It is associated with the human tumors Burkitt's lymphoma and anaplastic nasopharyngeal carcinoma [5]. The use of EBV for the generation of human B-lymphoblastoid cell lines by infection with EBV is a relatively straightforward technique which can have a success rate of over 95%. The cells are diploid when first transformed but become polyploid over long-term culture.

There are two critical steps in the procedure for establishing lymphoblastoid cell lines, namely: the condition of the blood prior to separation, and production of EBV supernatant B95-8 marmoset cell line (ECACC no. 85011419) with a sufficiently high virus titer. Fresh blood should be used, preferably within 24 h of sampling, but cell lines can be successfully established with blood which is up to 72 h old, as long as it has been kept at room temperature. Extremes of temperature (i.e. below 12°C or above 25°C) should be avoided as these will drastically reduce the transformation success rate. It is possible to freeze separated lymphocytes and store them in liquid nitrogen, prior to immortalization with EBV, without affecting the success rate.

Methods available

Spontaneous immortalization

Permanent cell lines can arise spontaneously (as a rare event) during the prolonged culture of primary cells. As the cells become senescent, foci of growing cells can be observed and occasionally permanent cell lines can be obtained. The main problem with this method is the lack of control by the investigator and the large number of cells likely to be needed before a line can be isolated.

Chemical transformation

The mouse L5178Y lymphoblast cell line was isolated after exposing cells to the chemical carcinogen methylcholanthrene [6]. This

References

1. Hayflick, L. and Moorhead, P.S. (1961) *Exp. Cell Res.* **25**, 585.
2. Huschtscha, L.I. and Holliday R. (1983) *J. Cell Sci.* **63**, 77.
3. Shay, J.W. and Wright, W.E. (1989) *Exp. Cell Res.* **184,** 109.
4. MacDonald, C. (1990) *CRC Crit. Rev. Biotechnol.* **10,** 155.
5. Epstein, M.A. and Achong, B.G. (1986) in *The Epstein Barr Virus*, pp. 2–11. William Heinemann, London.
6. Beer, J.Z., Budzicka, E., Niepokojczycka, E., Rosiek, O., Szumiel, I. and Walicka, M. (1983) *Cancer Res.* **43,** 4736.
7. Girardi, A.J., Jensen, F.C. and Koprowski, H. (1965) *J. Cell. Comp. Physiol.* **65**, 69.
8. Koehler, G. and Milstein, C. (1975) *Nature* **256,** 495.
9. Ruley, H.E. (1983) *Nature* **304,** 602.

method has been used to isolate many of the long-established cell lines, however, they have usually lost the differentiated characteristics of the tissue from which they were isolated.

Viral transformation

Infection of the diploid human line WI 26 with a DNA tumor virus, SV40, resulted in the production of the permanent cell line WI 26 VA4 [7]. Human lymphoblastoid cell lines are routinely produced by infecting peripheral human lymphocytes with EBV. This method works well with some cell types (e.g. human lymphoblasts), but there are safety implications if the cell line is capable of producing infectious virus.

Cell fusion

Another approach, to construct a fusion hybrid between the cell with a limited lifespan and a permanent cell line, has been particularly successful for generating hybridoma lines secreting antibody [8]. The two cell types are mixed together and treated with a fusogenic agent such as PEG. Some hybrids are quite unstable and rapidly lose chromosomes, and in many cases the hybrid has the phenotype of the undifferentiated permanent cell line [20].

Genetic modification

Cells can also be immortalized by using genetic engineering techniques to introduce and express viral genes which are capable of overcoming senescence. These genes have been categorized into two

10. Land, H., Parada, L.F. and Weinberg, R.A. (1983) *Nature* **304**, 596.
11. Rassoulzadegan, M., Cowie, A., Carr, A., Glaichenhaus, N., Kamen, R. and Cuzin, F. (1982) *Nature* **300**, 713.
12. Knudson, A.G. (1986) *Ann. Rev. Genet.* **20,** 231.
13. Bøyum, A. (1968) *Scand. J. Clin. Invest.* **21** (suppl. 97), 77.
14. Bøyum, A. (1964) *Nature* **204**, 793.
15. Doyle, A. (1989) in *Animal Cell Culture* (J.W. Poll and J.M. Walker, eds), pp. 43–47. Humana Press, Ottowa.
16. Bolton, B.J. and Spurr, N.K. (1996) in *Culture of Immortalized Cells* (R.I. Freshney and M.G. Freshney, eds), pp. 283–299. Wiley-Liss, New York.
17. Doyle, A. and Bolton, B.J. (1994) in *Basic Cell Culture: a Practical Approach* (J.M. Davis, ed.), pp. 243–270. IRL Press, Oxford.
18. Walls, E.V. and Crawford, D.H. (1987) in *Lymphocytes: a Practical Approach* (G.G.B. Klaus, ed.), pp. 149–162. IRL Press, Oxford.
19. Advisory Committee on Dangerous Pathogens. *Categorization of Pathogens According to Hazard and Categories of Containment* (ISBN 0/11/883761/3). HMSO, London.
20. McBride, B.W. (1993) in *Cell and Tissue Culture: Laboratory Procedures* (A. Doyle, J.B. Griffiths and D.G. Newell, eds), p. 2D:1. John Wiley & Sons, Chichester.

groups: those which are capable of immortalizing cells, that is establishing proliferating cultures from primary cells or increasing the lifespan of cells such as primary embryo fibroblasts; and those which can transform cells (e.g. the established mouse fibroblast cell line, NIH3T3 [9,10]). Thus the oncogenic transformation of rodent cells by DNA viruses occurs as a two-step process as a result of cooperation between two or more oncogenes [11].

Examples of oncogenes with immortalizing activity include the human adenovirus early region 1A (Ad E1A) and the polyoma large T-antigen coding region. Genes with transforming activity include polyoma middle T, the adenovirus early region 1B and the *ras* oncogenes. Genetic modification experiments are regulated. They may need prior approval but must certainly be notified to the appropriate authorities.

Separation of mononuclear cells from peripheral blood (see *Protocol 36*)
The method described below is essentially that of Bøyum [13,14] using a separation medium consisting of an aggregating agent, Ficoll, together with sodium metrizoate to give a density of 1.077 g/ml and osmolality of 280 mOsm/kg. The technique is a one-step centrifugation technique and isolates the mononuclear cells from whole blood.
Problems: If blood is older than 72 h or has been exposed to extreme temperatures then the separation does not work very well.

Protocols provided

Preparation of EBV-containing supernatant from the cell line B95-8 (see *Protocol 37*)

The marmoset cell line B95-8 is induced for the production of EBV, by reducing the temperature to 33°C and the serum content to 2%. It is essentially the same method as described in refs 15 and 16.

Problems: It is very important that a culture in the log phase of growth is used prior to reduction of serum and temperature, if not the cells may die too early without producing a high enough EBV titer.

Titration of EBV supernatant using cord blood mononuclear cells (see *Protocol 38*)

Outlined below is a method for evaluating the EBV supernatant produced in *Protocol 37*. This step can be omitted but it provides a way of comparing different batches of virus-containing supernatants.

Problems: It is not always easy to obtain cord mononuclear cells therefore, instead, the EBV supernatant should be tested out on control lymphocytes before being used on vital patient samples.

Establishment of human B-lymphoblastoid cell lines from peripheral blood mononuclear cells (see *Protocol 39*)

Protocol 39 is derived from the methods described in refs 15, 16 and 18, and from experience of generating over 15 000 cell lines using EBV. It involves infecting isolated lymphocytes with EBV, culturing them on 24-well plates in the presence of PHA and waiting for foci of immortalized B cells to grow out. The whole procedure takes 6–10 weeks.

Protocol 36. **Separation of mononuclear cells from peripheral blood.** B. Bolton

Reagents

Ficoll-Isopaque [e.g. Lymphoprep or Nycoprep 1.077 (Nycomed) or Accuspin Histopaque system (Sigma)]
Human blood sample ⚠
Lithium/sodium heparin, or ACD blood collection tubes
RPMI-1640 containing 10 units/ml heparin

Equipment

Bench-top centrifuge
15 ml Centrifuge tubes
Liquid nitrogen freezer
Rate-controlled freezer
Wide-bore Pasteur pipette plus tips

Procedure

1 Store the whole blood sample taken in lithium/sodium heparin, or ACD collection tubes at room temperature (15–25°C) prior to use. ①

2 Dilute the blood in RPMI-1640 containing 10 units/ml heparin at a ratio of 1:1. Addition of heparin is not necessary if blood is to be separated immediately. If the Sigma Accuspin system is used then there is no need to dilute the blood. ②

3 Carefully layer blood/RPMI-1640 mixture on to the separation medium at a ratio of 1 vol. separation medium to 2 vols diluted blood. This is usually achieved by adding 10 ml blood to 5 ml medium in a 15 ml centrifuge tube.

4 Centrifuge at 500 *g* for 20 min at room temperature.

Notes

This procedure takes approximately 2 hours for up to eight samples. If the Accuspin system is used then the process should only take 1.25 hours.

① EDTA-coated blood collection tubes may be used should those recommended be unavailable and should not cause any detrimental effect on EBV transformation.

② If large numbers of blood samples are to be processed then the Accuspin system (Sigma) is recommended. This has a filter part-way down the tube which prevents mixing of the two layers. Careful layering is not then required and the blood need not be diluted before use. The separation is as good as the traditional method described above, but

5 Carefully remove the mononuclear cell layer by using a wide-bore Pasteur pipette. Cells adhering to the walls of the tube can be dislodged by gentle scraping with the pipette tip. The tubes can be centrifuged for a further 10–20 min if a good separation has not been obtained. This is more often the case with blood samples separated after 72 h of sampling. ②③

6 Wash the cells by adding RPMI-1640 to a volume of 15 ml and mixing gently.

7 Centrifuge at 400 *g* for 10 min.

8 Check to see that a cell pellet has formed before discarding the supernatant. If a pellet is not visible repeat step 7. If the blood has a very low lymphocyte count then the pellet may only just be visible by eye. At this stage there may also be red cell contamination which should not affect the transformation process.

9 Resuspend the cell pellet in the same volume as in step 6 and repeat steps 6–8 twice. If desired, a viable cell count can be performed at this stage.

10 If cells are to be frozen, they should be resuspended in freeze media such as RPMI-1640 + 20% fetal bovine serum + 10% dimethyl sulfoxide, and frozen by rate-controlled freezing. The lymphocytes should then be stored until required, preferably in liquid or vapor phase nitrogen. However, they can be stored temporarily for up to 1 month at −80°C. If the cells are to be transformed immediately proceed to *Protocol 38*.

much quicker, however it can be more expensive when only a few samples are to be processed. Full instructions on its use are provided with the kit.

③ This method only works well with human blood.

Protocol 37. **Preparation of EBV-containing supernatant from the cell line B95-8.** B. Bolton

Reagents

B95-8 Marmoset cell line (ECACC or ATCC) ① ⚠
5% CO_2/95% air
2% Fetal bovine serum (FBS)
RPMI-1640

Equipment

Bench-top centrifuge
0.45 μm Filters (syringe or bottle-top type)
Flasks
37°C and 33°C Incubators
−80°C or Liquid nitrogen freezer
Pipettes, plus tips and pipettor
Vacuum pump
Vapor phase nitrogen freezer
Water bath at 37°C

Procedure

1 Maintain a culture of the B95-8 cell line in RPMI-1640 + 5% FBS at 3×10^5–9×10^5 cells/ml by diluting the cells 1:2–1:4 every 3–4 days.

2 When a sufficient number of cells has been obtained (depending on the amount of supernatant required) dilute the cells to 2×10^5 in RPMI-1640 + 2% FBS.

3 Gas the flasks with 5% CO_2, screw on the caps tightly and incubate at 33°C for 2–3 weeks without any changes of medium. ②

Notes

This procedure takes approximately 4–5 weeks.

① Unfortunately, many samples of the B95-8 cell line distributed informally between research laboratories are mycoplasma contaminated. It is therefore of the utmost importance that the basic starting material for the technique is obtained from a reliable source, such as an established culture collection which will guarantee pure authenticated and quality controlled cell lines.

4 On the day of harvest, up-end the flasks to allow cells to settle out.

5 Decant the supernatant into tubes and centrifuge at 400 *g* for 10 min at room temperature to remove any cell debris.

6 Filter the supernatant, through a 0.45 μm filter to remove cell debris and large particles, but not the virus. This is not intended as a sterilization procedure, as aseptic techniques should be followed throughout the process. ③

7 Aliquot the supernatant at required volumes (usually 1 ml) and store at −80°C or below, preferably in vapor-phase nitrogen. Freeze/thawing of the virus should be avoided as this will reduce the efficacy of EBV infection.

8 Prior to use, thaw at 37°C in a water bath and use immediately. If the virus is not to be used immediately, thaw at room temperature, keep on ice and use within 24 h.

② Some laboratories recommend the use of chemical inducing agents such as 12-*0*-tetradecanoyl phorbol-13-acetate (final concentration 20 ng/ml) or sodium-*n*-butyrate (3 mM) [18]. However, the titers produced by the above procedure are comparable with those obtained by chemical induction and the above chemical agents may interfere with B-lymphocyte activation by the virus.

③ EBV is an ACDP Category 2 pathogen [19].

Protocol 38. **Titration of EBV supernatant using cord blood mononuclear cells.**

B. Bolton

Reagents

CO_2 cylinder
Cord blood (5–10 ml) ①⚠
EBV supernatant⚠
10% Fetal bovine serum (FBS)
Growth medium
RPMI-1640

Equipment

Bench-top centrifuge
Centrifuge tubes
5% CO_2 incubator at 37°C
Culture flasks (plastic)
96-well Flat-bottomed microtiter plate
37°C Incubator
Inverted microscope
Pipette, plus tips and pipettor
Water bath at 37°C

Procedure

1 Separate the mononuclear cells from the cord blood as described previously (*Protocol 34*) above and resuspend in RPMI-1640 + 10% FBS at approximately 2×10^6–3×10^6 cells/ml. Use immediately if possible or culture overnight in a flask at 37°C. Cells give satisfactory results if cultured overnight but not if kept for longer. It is not advisable to pool donors. ①

Notes

This procedure takes approximately 4 hours unless the overnight incubation step is used. It will take up to 2 weeks to obtain a result.

① If it is not possible to obtain cord blood, the virus can be evaluated by testing its ability to immortalize lymphocytes from several normal donors.

2 Remove adherent cells by incubation in a plastic culture flask for 30 min. This step may be omitted if the cells have been cultured overnight.

3 Prepare serial 10-fold dilutions from neat to 1×10^{-6} of the virus supernatant to be tested in RPMI-1640 + 10% FBS. ②

4 Aliquot the cord mononuclear cells into centrifuge tubes to give 2×10^6 cells per 0.9 ml per tube.

5 Add 0.1 ml virus dilution to an aliquot of the above cells. As a negative control add 100 μl RPMI-1640 + 10% FBS to a tube of cells. A positive control can be set up using a previously prepared supernatant with a known positive titer.

6 Resuspend cells by tapping the base of each tube gently and incubate at 37°C for 1 h.

7 Add 1 ml of RPMI-1640 + 10% FBS to each tube and mix gently.

8 Plate cells out on to a 96-well flat-bottomed microtiter plate (200 μl per well). Use separate plates for each titration and two different batches of media when feeding.

9 Incubate plates at 37°C in 5% CO_2/95% air. Feed once a week by removing 100 μl medium without disturbing the cells and replacing with fresh medium. ③④

② These and all subsequent titrations should be carried out on ice. It is also important to change pipette tips between dilutions.

③ Successful immortalization of mononuclear cells can be assessed by examining the cultures using an inverted microscope. After 10–14 days, proliferating foci of B lymphocytes can be seen, which after about 28 days are usually visible macroscopically as large clumps of cells. The control cultures and cultures failing to be immortalized should contain only dying cells and debris.

④ The efficiency of immortalization is defined as $-\log_{10}$ of the virus dilution which induced 50% immortalization of the cultures. For routine use, the supernatant should have a titer of at least 10^{-3} and preferably 10^{-4}.

Protocol 39. **Establishment of human B-lymphoblastoid cell lines from peripheral blood mononuclear cells.** B. Bolton

Reagents

EBV supernatant (see *Protocol 37*) ⚠
Human lymphocytes ⚠
Mouse peritoneal macrophages (optional; see below).
Neomycin
Penicillin
Phytohemagglutinin solution (PHA) ⚠
RPMI-1640 + 20% fetal bovine serum (FBS)
Streptomycin

Equipment

Bench-top centrifuge
15 ml Centrifuge tubes
24-well Flat-bottomed culture dish
Gas boxes or CO_2 incubator
37°C Incubator
Liquid nitrogen freezer
Pipette, plus tips and pipettor
25 ml Tissue culture flasks
Water bath

Procedure

1 Thaw a vial of frozen lymphocytes (5×10^6–10×10^6 cells corresponding to lymphocytes yielded from a 10 ml blood sample) at 37°C and transfer to a 15 ml centrifuge tube. Fresh lymphocytes can also be used. Add pre-warmed RPMI-1640 + 10% FBS to a final volume of 15 ml and mix. Centrifuge at 150 *g* for 5 min. ①

Notes

This procedure takes approximately 3 hours for eight samples up to step 6. It may take up to 8 weeks before a final cell culture is established.

① The use of 24-well plates prepared with mouse peritoneal macrophages is recommended for low numbers of

2 Discard the supernatant and resuspend the cell pellet in 1 ml EBV supernatant.

3 Incubate at 37°C for 1–1.5 h, with gentle agitation halfway through incubation.

4 Centrifuge at 150 *g* for 10 min. Ensure that a pellet has formed before discarding the supernatant.

5 Resuspend cells in RPMI-1640 + 20% FBS, PHA (1% v/v), penicillin (100 units/ml), streptomycin (100 μg/ml), neomycin (50 μg/ml) to give 1×10^6–2×10^6 cells/ml. Pipette 1 ml of cells into each well of a 24-well flat-bottomed culture dish and incubate at 37°C in a 5% CO_2/95% air atmosphere. ②

6 The cultures should be fed twice weekly by removing half the supernatant and replacing it with fresh medium (no PHA added) without disturbing the cells. Ensure that pipette tips are not transferred between wells, to prevent cross-contamination.

7 After 1–2 weeks, the foci of B cells should be visible under an inverted microscope and the initial proliferation of T cells should regress.

8 The cultures may be expanded into 25 ml tissue culture flasks once established (usually after 2–3 weeks). This is often indicated by the media turning yellow overnight and lots of large clumps of cells present. The flasks should be kept upright and the cultures split 1:3 every 2–4 days depending on growth rate. ③④

peripheral blood lymphocytes (i.e. less than 1×10^6) [20], peripheral blood lymphocytes contaminated with red blood cells or peripheral blood lymphocytes isolated from old blood samples (over 3 days).

② Alternatively, cyclosporin A (toxic to T lymphocytes) may be used instead of PHA. However, it does not increase the efficiency of EBV transformation, and is more difficult to prepare. PHA results in the proliferation of T cells and has a mitogenic effect on B cells. In the author`s laboratory this has proved to be the most efficient technique available. One possible drawback is that mitogenesis leads to differentiation of committed B cells to plasma cells and then the loss of the C3 receptor, and thus the cells are not transformed.

③ During routine subculture of established lymphoblastoid cell lines it is important not to over-dilute the cells (i.e. split at ratios 1:3 or lower) as this can often halt their growth.

④ HEPES buffer may be added to the medium (final concentration 25 mM) if required to maintain the pH. This is particularly useful when dealing with large numbers of cell lines. If the culture is not growing well, growth supplements can be added at any stage of production of a cell line. A mixture of 2-mercaptoethanol (50 mM), nonessential amino acids (Life Technologies Ltd), sodium selenite (2×10^6 M), transferrin (4 mg/ml), sodium pyruvate (50 mM) used at a 1:50 dilution is recommended. This solution should be aliquoted and stored at −20ºC.

APPENDIX. SUPPLIERS

American Type Culture Collection, 12301 Parklawn Drive, Rockville, MD 20852-1776, USA. Tel. (301) 881 2600. Fax (301) 816 4361.

Ashby Scientific Ltd, Unit 11, Atlas Court, Hermitage Industrial Estate, Coalville, Leics LE67 3FL, UK. Tel. (01530) 832590. Fax (01530) 832591.

The Automation Partnership, Melbourn Science Park, Melbourn, Royston, Herts SG8 6HB, UK. Tel. (01763) 262026. Fax (01763) 262613.

BASF, ZHA A30, Ludwigstrasse, D-37056 Germany.

Becton-Dickinson (UK) Ltd, Between Towns Road, Cowley, Oxford OX4 3LY, UK. Tel. (01865) 748 844. Fax (01865) 717313.

Becton-Dickinson (Europe) Labware, 5 Chemin des Sources, 38241 Meylan, France. Tel. 04 7641 6464. Fax 04 7690 1965.

Becton-Dickinson (Immunocytometry Systems), 2350 Qume Drive, San Jose, CA 95131, USA. Tel. (408) 432 9475, (800) 223 8226. Fax (408) 954 2009.

Costar, 1 Alewife Centre, Cambridge, MA 02140, USA. Tel. (617) 868 6200, (800) 492 1110. Fax (617) 868 2076.

Corning Costar UK Ltd, 10 The Valley Centre, Gordon Road, High Wycombe, Bucks HP13 6EQ, UK. Tel. (01494) 484700. Fax (01494) 464891.

Dextran Products, PO Box 1360, Princeton, NJ 08542, USA.

Dow Chemical Co., Heathrow House, Bath Road, Houslow, TW5 9QY, UK. Tel. (0181) 848 8688. Fax (0181) 848 5400.

Expanded Metal Co. Ltd, PO Box 14, Stanton Works, Hartlepool, UK. Tel. (01429) 266633. Fax (01429) 866795.

Fison Scientific Equipment, Bishop Meadow Road, Loughborough, Leics LE11 0RG, UK. Tel. (01509) 231166. Fax (01509) 231893.

Heraeus Instruments GmbH, PO Box 1563, Hanau D-63405, Germany. Tel. 6181 35 300. Fax 6181 35 5973.

Becton-Dickinson Labware, Becton Drive, Franklin Lakes, NJ 07417-1886, USA. Tel. (201) 847 4222/800 343 2035. Fax 201 847 4882/800 847 2220.

Becton-Dickinson Microbiology Systems, 7 Loveton Circle, Sparks, MD 21152, USA. Tel. (410) 316 4000.

Bellco Glass Inc., PO Box B, 340 Edrudo Road, Vineland, NJ 08360, USA. Tel. (609) 691 1075. Fax (609) 691 3247. (UK distributor: Jencons Scientific Ltd.)

Bibby Sterilin Ltd, see Sterilin

Boehringer-Mannheim, Bell Lane, Lewes, East Sussex BN7 1LG, UK. Tel. (01273) 480444/(0800) 521578. Fax (01273) 480266/(0800) 181087.

Braun B., Biotech International GmbH, Schwarzenberge Weg 73–79, D-3508 Melsungen, Germany. Tel. (5661) 713704. Fax (5661) 713702.

Cellon, 22 rue Dernier, Sol L-2543, Luxembourg. Tel. 495 975.

Celltech (now Lonza Biologics plc), 228 Bath Road, Slough, Berks SL1 4DX, UK.

Heraeus Instruments Inc., 111A Corporate Boulevard, South Plainfield, NJ 07080, USA. Tel. (908) 754 0100, (800) 441 2554. Fax (908) 754 9494.

Heraeus Instruments Ltd, Wates Way, Brentwood, Essex CM15 9TB, UK. Tel. (01277) 231511. Fax (01277) 261856.

Imperial Laboratories (Europe) Ltd, West Portway, Andover SP10 3LF, Hants, UK. Tel. (01264) 333311. Fax (01264) 332412.

Jencons Scientific Ltd, Cherrycourt Way Industrial Estate, Stanbridge Road, Leighton Buzzard LU7 8UA, UK. Tel. (01525) 372010. Fax (01525) 379547.

Kirin, 1-17-1 Namamugi, Tsurumi-ku, Yokohama-shi, Konagawa 230, Japan.

Life Technologies Ltd, 3 Fountain Drive, Inchinnan Business Park, Paisley, Renfrewshire PA4 9RF, UK. Tel. (0141) 814 6100. Fax (0141) 814 6287. (Used to be Gibco.)

Lonza Biologics plc., 228 Bath Road, Slough, Berks SL1 4DY, UK. Tel. (01753) 777000. Fax (01753) 777001.

National Collection of Type Cultures, PHLS Central Public Health Laboratory, 61 Colindale Avenue, London NW9 5HT, UK. Tel (0181) 200 4400. Fax (0181) 200 7874.

New Brunswick Scientific UK Ltd, Edison House, 163 Dixons Hill Road, North Mimms, Hatfield AL9 7JE, UK. Tel. (01707) 275733. Fax (01707) 267859.

Nunc A/S, Kamstrupvej 90, Postbox 280, DK 4000, Roskilde, Denmark. Tel. 46 35 90 65. Fax 46 35 01 05. (UK suppliers: Life Technologies.)

Nycomed (UK) Ltd, Nycomed House, 2111 Coventry Road, Sheldon, Birmingham B26 3EA, UK. Tel. (0121) 742 2444. Fax (0121) 722 2190.

Oxoid Ltd, Wade Road, Basingstoke, Hants, UK. Tel. (01256) 841144. Fax (01256) 463388.

Pharmacia Biotech, 23 Grosvenor Road, St Albans, Herts AL1 3AW, UK. Tel. (01727) 814000/(0800) 318353. Fax (01727) 814001/(0800) 318354.

Pharmacia Biotech AB, Bjorkgatan 30, S-751 82 Uppsala, Sweden. Tel. 18 16 50 00. Fax 18 16 64 68.

Pharmacia Biotech (Biochrom) Ltd, 22 Cambridge Science Park, Milton Road, Cambridge CB4 4FJ, UK. Tel. (01223) 423723. Fax (01223) 420164.

Reactifs IBF, 35 Avenue Jean-Jaures, F-92390, Villeneuve-la-Garonne, France.

Sigma-Aldrich Chemical Co. Ltd, Fancy Road, Poole, Dorset BH12 4QH, UK. Tel. (01202) 733 114/(0800) 373731. Fax (01202) 715460/(0800) 378785.

Solihull Engineering, 1919 Green Road, Ann Arbor, MI 48105, USA.

Sterilin Ltd, 43–45 Broad Street, Teddington, Middlesex TW11 8QZ, UK.

TCS Biologicals Ltd, Botolph Claydon, Buckingham MK18 2LR, UK. Tel. (01296) 714071. Fax (01296) 715753.

Techmate Ltd, 10 Bridgetum Avenue, Old Wolverton, Milton Keynes MK12 5QL, UK. Tel. (01908) 322222. Fax (01908) 319941.

Techne (Cambridge) lTD, Duxford, Cambridge CB2 4PZ, UK. Tel. (01223) 372010. Fax (01223) 836838.

The Technology Partnership plc, Melbourn Science Park, Melbourn, Royston, Herts SG8 6EE, UK. Tel. (01763) 262626. Fax (01763) 261582.

Pharmacia Biotech Inc., 800 Centennial Avenue, PO Box 1327, Piscataway, NJ 08855-1327, USA. Tel. (800) 526 3593. Fax (800) 329 3593.

Planer Products Ltd, Windmill Road, Sunbury-on-Thames, TW16 7HD, UK. Tel. (01932) 779997. Fax (01932) 781151.

Ventrex Laboratories Inc., 217 Read Street, Portland, ME 04103, USA.

Wheaton Scientific, 1501 North 10th Street, Millville, NJ 08332, USA. Tel. (609) 825 1100. Fax (609) 825 1368. (UK distributor: Jencons Scientific Ltd.)

INDEX